Broadcast Indecency

Date Due

BRODART, CO.   Cat. No. 23-233-003   Printed In U.S.A.

Broadcast & Cable Series
Series Editor: Donald V. West, Editor/Senior Vice-President,
Broadcasting & Cable

Broadcast Indecency F.C.C. Regulation and the First Amendment

by
Jeremy Harris Lipschultz, Ph.D.

Focal Press
Boston   Oxford   Johannesburg   Melbourne   New Delhi   Singapore

Focal Press is an imprint of Butterworth-Heinemann.

Copyright © 1997 by Butterworth-Heinemann

❁ A member of the Reed Elsevier group

Library of Congress Cataloging-in-Publication Data

Lipschultz, Jeremy Harris, 1958–
    Broadcast indecency : F.C.C. regulation and the First Amendment /
by Jeremy Harris Lipschultz.
        p.   cm.
    Includes bibliographical references and index.
    ISBN 0-240-80208-X (pbk. : alk. paper)
    1. Broadcasting—Law and legislation—United States.   2. Obscenity
(Law)—United States.   3. United States. Federal Communications
Commission.   4. United States—Constitutional law—Amendments—1st.
I. Title.
KF2805.L57   1996
343.7309´94—dc20   96-20615
[347.303994]   CIP

British Library Cataloguing-in-Publication Data
A catalogue record for this book is available from the British Library.

The publisher offers special discounts on bulk orders of this book.
For information, please contact:

Manager of Special Sales
Butterworth-Heinemann
313 Washington Street
Newton, MA 02158-1626
Tel: 617-928-2500
Fax: 617-928-2620

For information on all Focal Press publications available, contact our World Wide Web home page at: http://www.bh.com/fp

10 9 8 7 6 5 4 3 2 1

Printed in the United States of America

Table of Contents

Preface

This book was conceived in early 1989. As a new faculty member at the University of Nebraska at Omaha, I wanted to know more about why broadcast indecency cases were beginning to surface. The insightful questions from students in my first class, as well as the questions from students since, prompted me to do research on the topic. I came to believe that broadcast indecency regulation by the Federal Communications Commission in the 1990s has become a defining issue for broadcasters and their relationship with the government. This book was written for broadcasters who need to know more about the law of broadcast indecency.

The earliest research for this book was warmly welcomed by the editors of *Communications and the Law* at a time when few saw the significance of the issues at stake. The University Committee on Research at UNO followed with funding for a content analysis reported in this book. There are many individuals who supported this project.

First, my family provided the most important encouragement: my wife, Sandy, our son, Jeff, and his grandmother Faye. We suffered through the loss of Sandy's father and both of my parents during these past few years, but knowing how they would have been interested in this book was and is a source of strength.

At UNO, the faculty of the Department of Communication have been extremely valuable in their knowledge of communication issues. Additionally, staff secretaries have been extremely helpful with computer work on the book. Finally, the administration—from department chairs to deans of the College of Arts and Sciences—have never waivered in their support of this research.

This book raises a lot of questions about the future of broadcast regulation in the United States and abroad. As with so many other areas now, broadcast indecency's future is difficult to predict. At best, we can make reasoned predictions based upon past and present conditions.

Jeremy Harris Lipschultz
Omaha, Nebraska
October 1995

Chapter 1

An Introduction to Issues in Broadcast Indecency

"As far as that sexually-oriented over-used 'F' word is concerned, I have heard it, used it, done it; I just think there are some places where it is improper, and even in some contexts repulsive."
FCC Commissioner James Quello

"We are dealing with protected speech. We are not dealing with obscenity, and we ought to leave that kind of decision to the parents of this country."
Timothy Dyk, partner, Jones, Day, Reavis & Poque

"To put the burden on parents and children and other people who have an interest in this is distortion of what the licensing procedure, I think, is all about."
Peggy Coleman, legal counsel, American Family Association

"The attempted repression of high literature (and free speech on public radio) is very much parallel with the mind repression we have seen with spiritual pollution campaigns in China, with the burning of degenerate works in Germany in the `30s, and with the attacks on bourgeois individualism and rootless cosmopolitanism and sexual license in literature in Russia. . . ."
Professor Allen Ginsberg, Brooklyn College

"[T]he American people are totally dissatisfied . . . with the moral quality, the excessive violence, the barnyard sex . . . that regularly goes over television and even so-called 'shock radio.'. . ."
Joseph Reilly, president, Morality in Media

When the Federal Communications Bar Association held a 1990 panel on "What's Indecent? Who Decides?" it was apparent that broadcasters arguing they were no longer "second-class citizens" with respect to the First Amendment were swimming upstream. Neither the public nor the legal rules were prepared to treat the words spoken on radio and television as falling under the umbrella of protection afforded the printed word.

The recent attempts by American government to regulate "indecent" speech on radio and television have placed the issue of broadcast indecency at the cutting edge of the battle to define the First Amendment. Regulators, if not telling broadcasters what to say or not say, are at least telling the people behind the microphones and cameras how to say it.

As the 1990s brought the Congress face-to-face with a marketplace oriented overhaul of sixty-year-old telecommunications regulations, some members continued to call for content rules that would regulate indecency and violence. At the same time in the courts, existing broadcast indecency rules were upheld. Broadcasters were "deeply disappointed," in the words of the National Association of Broadcasters, with a summer 1995 appeals court decision that supported the idea of a "safe harbor." Under these rules, broadcasters were free to air indecency after 10 P.M., but they were subject to punishment during the hours when audience sizes were the largest. While the *ACT III* decision upheld the constitutionality of Federal Communications Commission (FCC) policy, the *ACT IV* case that came next said the FCC was properly administering their rules.

Some argued that the narrowly focused attacks on the worst offenders help to create a "chilling effect" on the speech of all broadcasters in a variety of contexts. Howard Stern, the poster child of broadcast indecency, was labeled a "crude and obscene rabble-rouser" by an Ohio judge in 1995 who still conceded Stern had First Amendment rights of free speech—rights that made it illegal for a rival technician to cut his transmission line during a remote broadcast. Stern, meanwhile, published a second book. "We will get people reading books who are illiterate," he told the *New York Post*. Stern has been labeled a brilliant "social satirist" by his editor; the Federal Communications Commission has thought otherwise. In early September 1995 Infinity Broadcasting agreed to settle more than 100 claims by paying the government $1,715,000—"the largest settlement of its kind," according to the *Los Angeles Times*. The corporation agreed to pay the money to the U.S. Treasury and "agreed to alert all on-air personnel to the federal law prohibiting the broadcast of indecent speech. Infinity, which had vowed to fight the regulation on First Amendment grounds," apparently determined that a pragmatic stance made sense.

So-called content regulation of speech has historically been frowned upon by the courts, except where the government can make the case that there is a strong interest at stake. In the case of broadcast indecency, the interest at stake is argued to be the protection of children from potentially harmful messages. The courts, in turn, have struggled to balance the protection of children against the rights of adult listeners and viewers.

Previous authors have devoted small portions of communication law texts to the issue of broadcast indecency, even as they acknowledge it is an important issue. For example, Professor Don Pember (1993) wrote:

> The Federal Communications Commission has spent much of its time during the past twelve years cloaked in the mantle of the First Amendment, abandoning rules that restricted broadcasters' programming practices. But there is one exception: the agency has materially attempted to tighten the screws that limit the broadcast of obscene or indecent material. Section 1464 of the United States Code gives the FCC the power to revoke any broadcast license if the licensee transmits obscene or indecent material over the airwaves. . . . [H]aving the power to ban obscenity and indecency is one thing; defining obscenity and indecency is another.[1]

Without clear definitions of what is or is not indecent or obscene, broadcasters find themselves in a potential minefield of regulation. As we will see, a critical factor in the regulatory process is the *emergence of documented audience complaints*. This leaves open the possibility that a single listener or viewer, armed with time and taping equipment, can cause a lot of problems for the broadcaster who is walking near the edge of what is acceptable. While it "can be said with certainty that obscene materials have no First Amendment protection,"[2] broadcast "indecency . . . [has been] banned between the hours of 6 P.M. and 8 P.M."[3] T. Barton Carter, Marc Franklin, and Jay Wright (1993) treat broadcast indecency as "nonpolitical speech." While political speech is granted extraordinary protection from governmental control, Gillmor, Barron, Simon, and Terry (1990) conclude that nonpolitical speech may be banned:

> Most severely, the FCC can consider violations . . . when licenses are sought—either by renewal applications or by initial applications. Disqualification of a license application has been rare. . . . Instead,

the FCC has preferred to warn or admonish licensees, although occasionally there have been exceptions.[4]

The fact that the FCC took no action against any stations for indecency violations between the *Pacifica* decision, upholding regulation of the Carlin monologue's "Seven Dirty Words," in the mid-1970s and 1987 seems to suggest that there have been recent behavioral changes on the part of some broadcast licensees:[5] "The Commission's approach reflected, in part, satisfaction with industry self-regulation," according to Ginsburg, Botein, and Director (1991):

> Over time, however, a number of broadcasters—particularly in radio—became more aggressive (or less "discriminating," as one likes) in their choice of programming. Viewers had gained new access to sexually more explicit and more violent programming. . . . The result was a growing concern among large segments of the population that the mass media were contributing to a general decline in morality, particularly among younger, highly impressionable viewers. . . . As a result of these concerns, in the early to mid-1980s, Congress and the Reagan Administration began to express a desire for more active FCC enforcement. . . .

The result has been, according to Professor John Bittner (1994), that "[o]ne of the most complex areas of broadcast regulation is obscene, indecent, and profane material."[6] As a First Amendment matter, the broadcast media are distinguished from print and other forms. "Under the rationale that the electromagnetic spectrum is a limited [and scarce] resource, the FCC has the authority under the Communications Act to institute such rules as it deems in the public interest." The "linchpin of indecency enforcement is the protection of children from inappropriate broadcast material." The recent telecommunication reform legislation extended this type of thinking to the Internet, and to the issue of television violence.

The purpose of this book is to develop a comprehensive understanding of the significance of the broadcast indecency issue for broadcast professionals, policy makers, academics, students, and the general public. A synthesis of previous academic writing, court decisions, public communication, and FCC policy will show how attempts to regulate broadcast indecency—while founded on legitimate concerns for children in America—run the risk of deflating the value of free expression in a free society.

Attempts to Regulate Indecent Speech on U.S. Radio: Political Issues

Broadcast regulation in the United States, perhaps in every country, is a creature of the national political environment. Erwin Krasnow, Lawrence Longley, and Herbert Terry (1982), in the most widely recognized legal model, assert that the regulatory "process" is not a static structure:

> Despite persistent calls for an emphasis on the political aspects of policy making by agencies such as the FCC, most of the literature on broadcast regulation has emphasized instead such topics as the history and development of the FCC and the broadcast industry, the Commission's legal and administrative status, and the legal problems resulting from the combination of rule-making and adjudicative functions in the body. . . . Questions such as "Who gets what, when and how" from the process are rarely considered systematically. . . . Their answers are central to understanding the politics of broadcast regulation.[7]

The model identifies five "determiners" of policy: the FCC (a creature of a powerful Congress), the industry (which was involved in regulatory policy from the beginning), citizens' groups, the courts, and the White House. The analysis leads to seven generalizations about the process:

- Participants seek conflicting goals from the process
- Participants have limited resources insufficient to continually dominate the policy-making process
- Participants have unequal strengths in the struggle for control or influence
- The component subgroups of participant groups do not automatically agree on policy options
- The process tends toward policy progression by small or incremental steps rather than massive change
- Legal and ideological symbols play a significant role in the process
- The process is usually characterized by mutual accommodation among participants.

The case of broadcast indecency seems not to be an exception to these broad generalizations. For example, it is clear that public interest groups promoting "morality" on the airwaves have "conflicting goals" with broadcasters who find there is a sizable audience for borderline indecent material. In terms of time, money, and political resources, broadcast indecency is one issue among many to be dealt with by policy makers and those who wish to influence that policy. Content regulation today seems overshadowed by concerns about new information technologies and "convergence" or "blending" of traditional and nontraditional media.

Broadcasters continue to maintain a strong position in terms of a protective attitude on the part of the FCC. This may be because a segment of professional broadcasters has vocally denounced the likes of Howard Stern, the nation's most famous "shock jock," as bad for business. Further, indecency policy has "progressed" excruciatingly slow, and no "massive changes" are likely in the near future. The welfare of the nation's children has been used as a potent and powerful symbol by those wishing to enforce indecency law. And, ultimately, the FCC and the other participants have found room for "mutual accommodation" on the broadcast indecency issue. Matthew Spitzer's (1986) book *Seven Dirty Words and Six Other Stories: Controlling the Content of Print and Broadcast* argues that the government regulates broadcast content "more strictly than it regulates the printed word" for three reasons: "(1) to achieve economic efficiency, (2) to limit socially harmful conduct caused by people's exposure to sexually explicit or violent material, and (3) to prevent children's exposure to a variety of material that may harm them." He concludes, "[T]hese rationales cannot support the existing regulatory framework."[8]

It may be that the pluralistic, democratic model posited by Krasnow, Longley, and Terry functions because of the vagueness of the application of the guiding law and policy. With no clear standards, the political players are left to jockey for position.

Definitional Limitations of Indecency Law

The heart of broadcast indecency law is found in a federal statute, 18 United States Code, Section 1464:

> Whoever utters any obscene, indecent, or profane language by means of radio communication shall be fined not more than $10,000 or imprisoned not more than two years, or both.

At first, this law would seem to clearly establish criminal penalties for any use of obscene, indecent, or profane language by broadcasters. However, for a number of historical, legal, and political reasons, there has never been a direct test of the statute language, and it has not been applied literally.

Originally the indecency prohibition had been a part of the Communications Act, but it was moved to the U.S. Criminal Code in 1948.[9] Indecency prohibitions were part of the earliest Radio Act of 1927 and the Communications Act of 1934. Fines, possible imprisonment, and license suspension were threatened in the law. The 1948 revision, while moving the Section 1464 language from the Communications Act, left language in Sections 312 and 503 that granted the FCC authority to order a station off the air or impose fines of $2,000 per day per offense or deny license renewal to a licensee broadcasting obscene or indecent content. At the same time, however, the 1948 change distanced the indecency rules from Section 326 of the Communications Act—language that conflicts with indecency enforcement because of its anticensorship tone:

> Nothing in this Act shall be understood or construed to give the Commission the power of censorship over the radio communications or signals transmitted by any radio station, and no regulation or condition shall be promulgated or fixed by the Commission which shall interfere with the right of free speech by means of radio communication.

The anticensorship language, at one level, appears to be a strong statement of free expression consistent with the First Amendment of the United States Constitution: "Congress shall make no law . . . abridging the freedom of speech, or of the press. . . ." The courts, thus far, have dodged the sticky question of the potential political or First Amendment value of indecent speech. The FCC has attempted to treat indecency as a matter of context as well as content. Section 1464 lumps "obscene, indecent, or profane language," and this creates another problem; in the print media context, the law of obscenity has a long and distinct history. However, print and broadcast legal definitions continue to evolve. To understand broadcast indecency law, one must first review the historical origins in obscenity definitions.

Obscenity Law

Professor Don Pember notes that the Puritans were not the first to pass obscenity laws: "Many persons argue that pre-Revolutionary laws against blasphemy also prohibited obscenity, but the best available evidence doesn't support this thesis."[10] While Benjamin Franklin "had time to write erotic literature," his work apparently did not help define the law: The first recorded prosecution having to do with obscenity occurred in 1815, when a man named Jesse Sharpless was fined for exhibiting a painting of a man "in an imprudent posture with a woman." Earlier common law convictions apparently treated obscenity as "crimes against God."

Donald Gillmor and his coauthors cite Curl's case of 1727 as the one that brought obscenity into the common law of England: "[A] tasteless tract titled, in part, 'Venus in the Cloister or the Nun in Her Smock' was held by a court to jeopardize the general morality."[11] More than a century later the obscenity test became: "Whether the tendency of the matter charged as obscenity is to deprave and corrupt those whose minds are open to such immoral influence and into whose hands a publication of this sort may fall."[12] Later, America's obscenity statute used the Hicklin test. The test survived the first three decades of the twentieth century until courts began to judge entire works by how they would affect "the average reader."

The distribution of obscenity via the Post Office led to a series of cases ultimately producing the landmark *Roth* case in 1957; the Supreme Court upheld 5-4 an obscenity conviction and refused to grant First Amendment protection. The obscenity statute declared: "Every obscene, lewd, lascivious, or filthy book, pamphlet, picture, paper, letter, writing, print, or other publication of an indecent character . . ." was "nonmailable," and:

> Whoever knowingly deposits for mailing or delivery, anything declared by this section to be nonmailable, or knowingly takes the same from the mails for the purpose of circulating or disposing thereof, or of aiding in the circulation or disposing thereof, shall be fined not more than $5,000 or imprisoned not more than five years, or both.[13]

The Supreme Court's interpretation of the statute is of interest in the broadcast indecency area because it strikes to the core of questions over content regulation. The case is abstracted below.

ROTH v. UNITED STATES

354 U.S. 476, 77 S.CT. 1304, 1 L.ED. 2D 1498 (1957)

Mr. Justice Brennan delivered the opinion of the Court . . .

The guaranties of freedom of expression in effect in 10 of the 14 States which by 1792 had ratified the Constitution, gave no absolute protection for every utterance. Thirteen of the 14 States provided for the prosecution of libel, and all those States made either blasphemy or profanity, or both, statutory crimes. As early as 1712, Massachusetts made it criminal to publish "any filthy, obscene, or profane song, pamphlet, libel, or mock sermon" in imitation or mimicking of religious services. . . . Thus, profanity and obscenity were related offenses. . . .

The protection given speech and press was fashioned to assure unfettered interchange of ideas for the bringing about of political and social changes desired by the people. . . . All ideas having even the slightest redeeming social importance—unorthodox ideas, controversial ideas, even ideas hateful of the prevailing climate of opinion—have the full protection of the guaranties, unless excludable because they encroach upon the limited area of more important interests. But implicit in this history of the First Amendment is the rejection of obscenity as utterly without redeeming social importance. This rejection for that reason is mirrored in the universal judgment that obscenity should be restrained, reflected in the international agreement of over 50 nations, in the obscenity laws of 48 States, and in the 20 obscenity laws enacted by Congress from 1842 to 1956. This is the same judgment expressed by this Court in *Chaplinsky v. New Hampshire*, 315 U.S. 568, 571–572:

> "There are certain well-defined and narrowly limited classes of speech, the prevention and punishment of which have never been thought to raise any Constitutional problem. These include the lewd and obscene. . . . It has been well observed that such utterances are no essential part of any exposition of ideas, and are of such slight social value as a step to truth that any benefit that may be derived from them is clearly outweighed by the social interest in order and morality. . . ."

We hold that obscenity is not within the area of constitutionally protected speech or press.

However, sex and obscenity are not synonymous. Obscene material is material which deals with sex in a manner appealing to prurient interest. The portrayal of sex, e.g., in art, literature, and scientific works, is not itself sufficient reason to deny material the constitutional protection of freedom of speech and press.

Sex, a great and mysterious force in human life, has indisputably been a subject of absorbing interest to mankind through the ages; it is one of the vital problems of human interest and concern. . . .

The early leading standard of obscenity allowed material to be judged merely by the effect of an isolated excerpt upon particularly susceptible persons. . . . Some American courts adopted this standard but later decisions have rejected it and substituted this test: whether to the average person, *applying contemporary community standards, the dominant theme of the material taken as a whole appeals to prurient interest* (emphasis added). The Hicklin test, judging obscenity by the effect of isolated passages upon the most susceptible persons, might well encompass material legitimately treating with sex, and so it must be rejected as unconstitutionally restrictive of the freedoms of speech and press. . . .

The Judgments are Affirmed.

Mr. Chief Justice Warren, concurring in the result.

It is not the book that is on trial; it is the person. The conduct of the defendant is the central issue, not the obscenity of a book or picture. The nature of the materials is, of course, relevant as an attribute of the defendant's conduct, but the materials are thus placed in context from which they draw color and character. A wholly different result might be reached in a different setting. . . .

Mr. Justice Harlan, concurring (in part) . . . and dissenting (in part) . . . The Court seems to assume that "obscenity" is a peculiar genus of "speech and press," which is as distinct, recognizable, and classifiable as poison ivy is among other plants. . . .

Every communication has an individuality and "value" of its own. The suppression of a particular writing or other tangible form of expression is, therefore, an individual matter, and in the nature of things every such suppression raises an individual constitutional problem, in which a reviewing court must determine for itself whether the attacked expression is suppressible within constitutional standards. Since those standards do not readily lend themselves to generalized definitions, the constitutional problem in the last analysis becomes one of particularized judgments which appellate courts must make for themselves.

Mr. Justice Douglas, with whom Mr. Justice Black concurs, dissenting.
When we sustain these convictions, we make the legality of a publication turn on the purity of thought which a book or tract instills in the mind of the reader. I do not think we can approve that standard and be faithful to the command of the First Amendment, which by its terms is a restraint on Congress and which by the Fourteenth is a restraint on the States.

By these standards punishment is inflicted for thoughts provoked, not for over acts or anti-social conduct. . . .
The tests by which these convictions were obtained require only the arousing of sexual thoughts. Yet the arousing of sexual thoughts and desires happens every day in normal life in dozens of ways. . . .
The test of obscenity the Court endorses today gives the censor free range over a vast domain. To allow the State to step in and punish mere speech or publication that the judge or jury thinks has an undesirable impact on thoughts but that is not shown to be a part of unlawful action is drastically to curtail the First Amendment. As recently stated . . . "The danger of influencing a change in the current moral standards of the community, or of shocking or offending readers, or of stimulating sex thoughts or desires apart from objective conduct, can never justify the losses to society that result from interference with literary freedom."

I can understand (and at times even sympathize) with programs of civic groups and church groups to protect and defend the existing moral standards of the community. . . . When speech alone is involved, I do not think the government, consistently with the First Amendment, can become sponsor of any of these movements. I do not think that government, consistently with the First Amendment, can throw its weight behind one school or another. Government should be concerned with antisocial conduct, not with utterances. Thus, if the First Amendment guarantee of freedom of speech and press is to mean anything in this field, it must allow protests even against the moral code that the standard of the day sets for the community.

I would give the broad sweep of the First Amendment full support. I have the same confidence in the ability of our people to reject noxious literature as I have in their capacity to sort out the true from the false in theology, economics, politics, or any other field.

Ultimately, Justice Douglas placed absolute faith in the "marketplace of ideas."[14] The free flow of ideas, however, has been weighed against the idea that mass communicators need to exercise "social responsibility."[15]

Theodore Peterson identified six functions of the press under social responsibility theory:

1.   servicing the political system by providing information, discussion, and debate on public affairs;

2.   enlightening the public so as to make it capable of self-government;

3.   safeguarding the rights of the individual by serving as a watchdog against government;

4.   servicing the economic system, primarily by bringing together the buyers and sellers of goods and services through the medium of advertising;

5.   providing entertainment;

6.   maintaining its own financial self-sufficiency so as to be free from the pressures of special interests.

While some who have challenged the FCC on the broadcast indecency issue argued that their content serves the first three functions, it seems a stretch to make that case. It is easy, however, to make the case that broadcast indecency serves the last three functions in the model. High ratings for such personalities as Howard Stern clearly show economic and financial value in the programming. Still, on theoretical grounds, broadcast indecency is more easily defended from an absolutist (literalist), marketplace-of-ideas interpretation of the First Amendment.

Fred Siebert grounded the free exchange of ideas in individual "liberalism" and the rationality of people: "Although men frequently do not exercise their God-given powers of reason in solving human problems, in the long run they tend, by the aggregate of their individual decisions, to advance the cause of civilization."[16] The view is: "Man differs from lower animals in his ability to think, to remember, to utilize his experience, and to arrive at conclusions." Siebert's application of liberalism to the press became his interpretation of Milton: "Let all with something to say be free to express themselves."[17] The distasteful presumably would be purged in the "self-righting" process: "Ultimately the public could be trusted to digest the whole, to discard that not in the public interest and to accept that which served the needs of the individual and of the society which he is a part."[18]

Professor Siebert found the chief instrument of control to be the judicial system:

> In the United States the courts are paramount since they not only apply the law of the land to the press but also determine when the other branches of government are overstepping their authority in imposing restrictions which might contravene constitutional protections. In the last analysis, under our constitutional system the courts determine the limits to which government may go in exercising the authority over the mass media. In other democratic countries, tradition or the legislature performs this function.[19]

While the liberal model creates a stable framework for the self-righting process to take place, Siebert acknowledged that "prohibitions against the dissemination of obscene and indecent materials" was a "commonly accepted restraint" on mass media:

> No sound basic principles have been developed to support the laws against obscenity other than that such restraints are necessary to protect morality. Morality itself is difficult to define, and both courts and legislatures have struggled for several centuries to arrive at an acceptable definition of obscenity. The definition of obscenity has usually been determined by an aggressive minority or by some judge's estimate of the current state of morality. Although some libertarians argue against all types of control based on obscenity, the majority agree that the state has an obligation to protect society, or at least some parts of it, from lewd and indecent publications.[20]

In the case of broadcast indecency, we will explore in this book the issue of protection of children from the presumed harmful effects of broadcast indecency. It is an argument that draws heavily from research on the effects of obscenity and pornography.

Most significantly in the study of broadcast indecency, the American law of obscenity was ultimately set in the *Miller v California* (1973) definition:

> (a) whether "the average person, applying contemporary community standards" would find the work, taken as a whole, to appeal to the prurient interest; (b) whether the work depicts or describes, in a patently offensive way, sexual conduct specifically defined by the applicable state law; and (c) whether the work, taken as a whole, lacks serious literary, artistic, political, or scientific value.[21]

The definition, as we shall see, is at the core of how we also define broadcast indecency.

Indecency Legal Definitions

The courts—most notably, the United States Court of Appeals, District of Columbia circuit three times in four years—relied upon the FCC's indecency definition:

> The Commission interpreted the "concept of 'indecent' [to be] intimately connected with *the exposure of children to language* that describes, in terms patently offensive as measured by contemporary community standards for the broadcast medium (emphasis added), sexual or excretory activities and organs, at times of the day when there is a reasonable risk that children may be in the audience.[22]

The American law of broadcast indecency was cast by the U.S. Supreme Court in the 1970s *Pacifica* ruling. A majority of the justices agreed with the FCC that indecency should be broadcast only during late night, overnight hours.

The standard is broader than that for obscenity in the print media context. While it singles out broadcasters, it goes well beyond discussion of sexual topics and factors in the time of day of the broadcast.

The *Pacifica* decision by the U.S. Supreme Court in 1978, which will be discussed in detail in this book, sanctioned the FCC regulation of George Carlin's "Filthy Words" monologue broadcast at 2 P.M. in New York. For a decade following that decision, the FCC's action was limited to station broadcasts before 10 P.M. involving "repeated use, for shock value, or words similar or identical to those satirized in the Carlin 'Filthy Words' monologue," namely: "shit, piss, fuck, cunt, cocksucker, motherfucker, and tits."[23]

By 1987, however, the FCC had broadened its regulation beyond the seven words and also attempted to close the "safe harbor" to midnight to 6 A.M.—a move rejected by the federal courts: the FCC "defines broadcast indecency as language or material that, in context, depicts or describes, in terms patently offensive as measured by contemporary community standards for the broadcast medium, sexual or excretory activities or organs."[24] While the courts have rejected a narrowed safe harbor, they have not rejected the FCC's "avowed objective" in indecency enforcement: "not to establish itself as a censor but to assist parents in controlling the material young children will hear."[25] The "safe harbor" idea of "chanelling" indecent broadcasts to late-night hours, when children are not as likely to be in the audience, remains FCC policy today.

Concern over what children will hear and see from the mass media is not a new concern. Broadcast indecency can be placed within the larger context of the public's relationship with the instruments of mass communication. We are prone to both marvel at the potential of new and old technology and fear its possible damaging effects, as we will see below.

Broadcast Indecency as Social Phenomenon— Mass Communication as Theory and Model

Mass communication theories and models, grounded in concern over media effects, illustrate how the case of broadcast indecency is part of a larger social phenomenon. The rapid rise of the movie industry in the 1920s immediately raised questions about the effects of the movies on the nation's children. In 1929, an estimated 40 million of the 90 million moviegoers were minors, and 17 million were under the age of fourteen.[26]

A content analysis of 1500 films of the time, while grouping them into ten types, found only three thematic categories accounted for three-fourths of the subject matter: love, crime, and sex. The Payne Fund studies, which attempted to determine the effects of the movies on children, found: (1) high numbers of young children in the audience, (2) unusually high levels of memory of what they had seen at the movies, (3) a view of morality in the movie content different from that of the audience viewing, (4) and potential behavioral effects:

> The investigators found that movie "fans" were usually rated lower in deportment by their teachers than those children who did not attend the movies frequently. The fans also had less positive reputations; they did worse in their academic work, and they were not as popular with their classmates as the comparison group.

Further, a study by sociologist Herbert Blumer suggested that movie content was a substantial influence: "The subjects reported that they had imitated the movie characters openly in beautification, mannerism, and attempts at lovemaking."

While one would not want to be quick to jump at the effects conclusions of that day, the fears of the 1920s are an important historical beginning for us to understand how today's media critics attack broadcast indecency. Consider, for example, the statement in 1994 by longtime FCC Commissioner James Quello; it links the need to protect children from "excessive, explicit, deviant sex" to a call for stronger regulation of licensees.

Statement of FCC Commissioner James H. Quello, 1994 FCC LEXIS 1110, March 17, 1994

And now for the subject I'm not supposed to talk about even though everyone else does—Infinity Broadcasting and Howard Stern. I am informed that there were new complaints against Howard Stern filed since my dissent to the transfer of KRTH(FM) was released February 1, 1994.

At that time I wrote: "The pattern of egregious repeated violations of FCC indecency rules is so flagrantly aggravated by six new complaints against Infinity and Howard Stern that I am impelled to dissent to being a party to any action that would result in approving additional stations for Infinity. . . . Once the issues raised by the *ACT III(a)* court case have been fully addressed, the full Commission should reconsider Infinity's repeated flaunting of the FCC indecency rules which could bear on the fitness of the licensee. It is apparent that previous FCC fines have not had a deterrent effect. Additional fines could merely be written off as a cost of doing business. In fact, no fines have been paid. Infinity is actively opposing the FCC fines exercising its legal rights. . . ."

It is unlikely and counter-intuitive to believe decent, responsible people would find it in the public interest to support additional outlets for licensees propagating lewdness, incest, and deviant behavior demeaning to women, blacks and gays. Certainly, First Amendment rights were not conferred by our Founding Fathers for repulsive, indecent or possibly obscene purposes. It must surely be embarrassing for First Amendment absolutists to defend language quoted in my dissent, broadcast at times children and young people are in the audience.

In this case, substantial justice should ultimately prevail over any technical legalism that First Amendment purists may utilize. Substantial justice requires the FCC to serve overall public interest. Today there is an overwhelming public outcry against excessive, explicit, deviant sex and glamorized violence and brutality on the air. It requires responsible action by the FCC and by public service minded broadcasters.

Today there is considerable public support for "three strikes and you're out" (lifetime penalty) for unlawful conduct. In each previous case in fining Infinity, the Commission has stated that it would consider further actions should Infinity broadcast indecent material in the future. Each time there has been yet another violation.

How many "next times" can the Commission tolerate? At some point, common sense alone would dictate that it is obvious the fines have not had a deterrent effect.

As a former newsman and broadcaster I consider myself a strong advocate of First Amendment rights. I believe reporters and broadcasters have a right to be wrong; a right to be insufferable smart-asses; but not a right to violate established indecency and obscenity laws.

Quello's call for social responsibility on the part of broadcasters operating under the public interest standard, and for stronger regulatory enforcement of indecency and obscenity laws, is in stark contrast to the libertarian ideal expressed by Justice Douglas in his *Roth* dissent.

While much of the modern media effects research centers on television violence rather than sexual content, and while the results are contradictory, most would accept the proposition that at the cultural level our mass media content influences our notions of acceptable behavior.[27] The mass media may enforce social norms as a result of "exposure" to conditions that deviate from professed public morality. Another consequence of the mass media is a "narcotizing" of the average reader or listener as a result of the flood of media stimuli. Mass media are among the most respectable and efficient of social narcotics, and increasing dosages may be transforming our energies from active participation to passive knowledge.[28]

The public debate over media and public versus private morality is not limited to the United States. According to mass media scholar Denis McQuail:

> There has been continued debate in many countries over . . . morals, decency and portrayals of matters to do with pornographic sex, crime and violence. While direct censorship and legal limitations have diminished in proportion to more relaxed standards in most societies, there remain limits to media freedom on grounds of the protection of minors from undesirable influences. . . . The issue has become further complicated by similar claims on behalf of women, who may either be portrayed in degrading circumstances or risk becoming the object of media-induced pornographic violence.[29]

While media organizations may be driven by commercial pressures, there is considerable evidence in the research on media effects that audiences actively employ "expectation and judgment" when using mass media. That is: "The values most frequently expressed about content are rather familiar and often stem from traditional judgments embedded in the culture and handed on mainly by the institutions of education, family and religion," McQuail writes. "They seem to favor the informational, educational or moral over entertainment and popularity." Part of the concern over broadcast indecency, however, clearly stems from a suspicion that American communities are struggling with weaker institutions of education, family, and religion, and the result may be that the mass media institution is becoming more influential by filling a cultural void. In short, people worry that for some young boys, Howard Stern may be a stronger male influence than their own absentee fathers.

Social Reality and Social Construction through Mass Media

Shock jock Howard Stern in the 1990s has become a media icon. His campaign for governor of New York on the Libertarian Party ticket included a mocking campaign organization: "Joining the entourage was a scantily clad woman with large breasts and a lavishly tattooed lesbian who claims to have had sex with space aliens."[30] In 1994 when his controversial morning radio program was scheduled to be added to cable television across the country, he said: "That's what's so wrong with America, that even a dope like me can realize his dreams."[31] The author of two best-selling autobiographies, Stern has a caustic approach to mass communication that, for many, runs counter to the moral mainstream. It is clear, however, that the publicity Stern has received in the broader mass media of newspapers and television has had an effect on the regulatory process. Stern and Infinity Broadcasting became a lightning rod in the broadcast indecency debate—a phenomenon that may have been "grounded" by the settlement in late 1995.

Road Map for Study: Concepts, Histories, Theories, Effects, and Economies

This book treats broadcast indecency as more than a simple regulatory problem in American law. The approach cuts across legal, social, and economic concerns. The author takes the view that media law and regulation cannot be seen within a vacuum that ignores social and cultural realities.

As such, we will first consider conceptual problems in the application of broadcast indecency law by the FCC. We will revisit the seminal *Pacifica* decision. Following that exploration, we will return to the origins of obscenity by reviewing the *Miller v California* and *Pope v Illinois* cases.

It will also be necessary to explore mass communication theory from a gender-based approach. Gender studies on the use of humor and language in American society are crucial in the understanding of the meaning of broadcast indecency. This theoretical base will be useful in the case study of so-called nonactionable complaints against broadcasters—complaints reviewed but rejected by the FCC.

In another case study, we will review how the role of a station's audience and community may influence the outcome of a broadcast indecency case. Additionally, the approach that individual broadcast managers take can be influential in the interaction with FCC regulators. The recent decision in the *Branton* U.S. Court of Appeals case will be discussed in connection with indecent news broadcasts and legal standing by audience members to complain.

Howard Stern and other radio shock jocks will be discussed in terms of their political and social significance. Indecent communicators will be categorized, and listeners analyzed through their communication. Ultimately, we will return to the question of the media effects of broadcast indecency, particularly effects on children—the group the government says it wants to protect.

Three interrelated issues will conclude the discussion of broadcast indecency in this book: the role of ratings, advertising, and profit; the impact of an emerging international media context; and the significance of new technologies. The disjointed policy of broadcast indecency will be seen as functional in the progression of commercial mass media.

Three models for regulation of speech can be considered in any call for review of mass media content such as broadcast indecency. There are three basic positions:

1. the case for more regulation,
2. the case for the status quo, and
3. the case for "more speech" as the solution.

In case #1, we saw how Commissioner Quello felt that because Infinity could not follow the law of indecency, the FCC needed tighter enforcement. In case #2, industry representatives argue against regulatory change because the current system has been effective in promoting development. In case #3, First Amendment absolutists argue, as we saw in Justice Douglas's Roth dissent, that good will ultimately prevail. If you do not like what is being broadcast, the solution is to provide an alternative that is commercially successful in the marketplace. Which position do you support? Is it possible to construct an argument that uses elements of all three positions?

Source: Adapted from National Issues Forums Institute, Public Agenda Foundation, *The Boundaries of Free Speech: How Free Is Too Free?* Dubuque, Iowa: Kendall/Hunt Publishing, 1991.

At the end of each chapter, the reader will find a "Manager's Summary," which will summarize the key points as applied to a broadcaster's situation.

Manager's Summary

Broadcast Indecency Overview

As a legal matter:

- Broadcast indecency is not fully protected speech under the First Amendment.
- Broadcast indecency is historically linked to concerns over obscenity.
- Desires to serve the marketplace should consider the legal requirement to serve community interests.
- Blatantly thumbing one's nose at the regulations is bound to lead to future legal troubles—particularly if interests in your community are willing to file complaints. Be prepared to pay fines—and to tarnish your licensee record.

As a social matter:

- It is prudent to assess the narrow interests of your audience and compare them to the larger community interests.
- Your station should address its own view of "social responsibility" to children that may be present in the audience. How your talent say things, in the eyes of the community, may be as important as what they say.

Notes

1. Don R. Pember, *Mass Media Law*, 6th ed. Dubuque, Iowa: Brown & Benchmark, 1993, p. 571.
2. Kenneth C. Creech, *Electronic Media Law and Regulation*. Boston, Mass.: Focal Press, 1993, p. 120.
3. Ibid., p. 127.
4. Donald M. Gillmor, Jerome A. Barron, Todd F. Simon, and Herbert A. Terry, *Mass Communication Law: Cases and Comment*, 5th ed. St. Paul, Minn.: West Publishing, 1990, p. 819.

5. Douglas H. Ginsburg, Michael H. Botein, and Mark D. Director, *Regulation of the Electronic Mass Media: Law and Policy for Radio, Television, Cable and the New Video Technologies*, 2nd ed. St. Paul, Minn.: West Publishing, 1991, pp. 538–539.

6. John R. Bittner, *Law and Regulation of Electronic Media*, 2nd ed. Englewood Cliffs, New Jersey: Prentice Hall, 1994, p. 122, 126 citing, In the Matter of Liability of Sagittarius Broadcasting Corporation, 1992 FCC LEXIS 6042, (October 23, 1992), citing Action for *Children's Television v FCC*, 271 U.S. App. D.C. 365, 852 F.2d 1332, 1340 (D.C. 1988).

7. Erwin G. Krasnow, Lawrence D. Longley, and Herbert A. Terry, *The Politics of Broadcast Regulation*, 3rd ed. New York: St. Martin's Press, 1982, pp. 2, 33–132, 278–283.

8. Matthew L. Spitzer, *Seven Dirty Words and Six Other Stories: Controlling the Content of Print and Broadcast*. New Haven, Conn.: Yale University Press, 1986, p. 1.

9. Creech, p. 120.

10. Pember, p. 414.

11. Gillmor, Barron, Simon, and Terry, p. 647, citing 2 Strange 788, 93 Eng.Rep. 849 (K.B. 1727).

12. Ibid., citing *R. v Hicklin*, L.R. 3 Q.B. 360 (1868): "It seemed consistent with the Tariff Act of 1842, our first obscenity law, prohibiting transportation into the United States of obscene literature, and with other laws that were defining freedom of speech as freedom for 'clean' speech only."

13. 18 U.S.C., Section 1461.

14. John Milton's often quoted words in Areopagitica (1644) were: "And though all the winds of doctrine were let loose to play upon earth, so Truth be in the field, we do injuriously by licensing and prohibiting to misdoubt her strength. Let her and Falsehood grapple; who ever knew Truth put to the worse, in a free and open encounter?" Justice Holmes brought life to this as American legal concept in *Abrams v United States*, 250 U.S. 616 (1919).

15. Fred S. Siebert, Theodore Peterson, and Wilbur Schramm, *Social Responsibility and Soviet Communist Concepts of What the Press Should Be and Do*. Urbana: University of Illinois Press, 1963, p. 74.

16. Ibid., p. 40.

17. Ibid., p. 45

18. Ibid., p. 51.

19. Ibid., p. 53.

20. Ibid., p. 55.

21. *Miller v. California*, 413 U.S. 15 (1973).

22. *Action for Children's Television v FCC (ACT IIIa)*, 11 F.3d 170 (D.C. Cir. 1993), at 172, quoting *Pacifica*, 438 U.S. at 731–32.

23. In re Infinity Broadcasting Corp. of Pa., 3 F.C.C.R. 930 (1987), and quoting the Carlin words from the *Pacifica* decision.

24. *ACT IIIa*, at 172, quoting 1993 Order, 8 F.C.C.R., at 704–705, note 8.

25. *ACT IIIa*, quoting *ACT I*, 852 F.2d, at 1334 (emphasis in original).

26. Shearon A. Lowery and Melvin L. DeFluer, *Milestones in Mass Communication*, 2nd ed. New York: Longman, 1988, pp. 33–38.

27. Werner J. Severin, and James W. Tankard, Jr., *Communication Theories: Origins, Methods, and Uses in the Mass Media*, 3rd ed. New York: Longman, 1992, pp. 255–301.

28. Denis McQuail, *Mass Communication Theory, an Introduction*, 3rd ed. London: Sage, 1994, p. 138.
29. Ibid., p. 311.
30. William F. Buckley Jr., "Howard Stern Stumps as Libertarian," Universal Press Syndicate, *Omaha World-Herald*, 1 June 1994, p. 11.
31. "Stern radio show to be on cable TV," Associated Press, *Omaha World-Herald*, 1 June 1994, p. 40.

Chapter 2

Conceptual Problems of Policy and Application

Federal regulation of broadcast "indecency" has sparked academic and professional debate, especially during the past two decades.[1] Armed with substantial regulatory authority delegated from the Congress and upheld in the courts, the Federal Communications Commission developed policy to restrict certain forms of nonpolitical speech. The early 1970s were years in which the FCC found itself reviewing content of radio programs—including the speech of announcers, music lyrics, and audience members.[2]

Broadcast policy evolved concurrently as the nation's highest court considered the degree of constitutional protection for speech, generally, in the *Roth* and the *Miller* cases. By the time of the *Pacifica* decision, dealing directly with over-the-air indecency, the FCC itself had begun to develop a philosophy which argued for a need to protect children.

Language in the Communications Act of 1934 and in a federal criminal statute (18 U.S.C. 1464) provided the FCC with an apparent right to penalize, if not forbid, obscene and indecent broadcasts. The *Pacifica* court had concluded: "[T]he validity of civil sanctions is not linked to the validity of the criminal penalty."[3]

Whether the offensive words were in a poem,[4] a taped interview with a rock musician,[5] or "topless" radio talk telephone calls, the FCC had begun to take the view that some, but not all, discussion of sex was improper.[6] In particular, use of specific words or innuendo could place the broadcast licensee under FCC review.

Because *existing broadcasters hold an expectancy of license renewal*, holding a renewal for review by a hearing examiner is considered a serious penalty in its own right.

The *Pacifica* decision, in which the U.S. Supreme Court supported the right of the FCC to channel indecency to late-night hours, failed to solve the problematic nature of the regulation. FCC regulators were faced with a case-by-case approach to enforcement.[7] Ater a brief lull following the U.S. Supreme Court decision, broadcasters—radio stations, in particular—returned with new formats that challenged the ability of the FCC to define and identify indecent programming.[8]

This chapter traces the history of the broadcast indecency issue to a recent line of cases. The goal is to analyze action by the FCC and identify underlying principles. Following this review, we will revisit the landmark *Pacifica* decision to see its significance.

The FCC and Broadcast Indecency: Key Questions

1. What generalizations can be derived from recent FCC indecency action against radio stations?
2. Have new regulatory principles emerged since the *Pacifica* decision? If so, what are they?
3. Is it possible to define and implement a coherent policy on the regulation of indecent language? If so, how?
4. What social values justify the need to regulate broadcast indecency?
5. Is the legal authority to regulate indecency a sound one? If so, are clarifications needed? If not, where are the weaknesses?

It is argued here that legal content definitions are problematic. Debate over restriction of "indecent" content in broadcasting—whether that be over-the-air, cable, or newer methods—continued in the 1980s[9] with little attention paid to two important issues: (1) the failure to define indecency or even clarify a distinction with obscenity that does not fall under the umbrella of constitutional protection, and (2) the nearly unquestioning assumption that there is a strong interest in the "protection" of children from "indecent" broadcast content. Most often, it is explicit profanity or sexual description that concerns the FCC, yet we know little from research about the effects of such expression on children.

Professor Howard Kleiman has found three legal dimensions to the issue of indecent cable television broadcasts: (1) the privacy argument that individuals have a "right" to keep their homes free of invasion by indecent cable programming, (2) the strong interest in the protection of children, and (3) the application of the scarcity rationale to cable content regulation by government entities.[10] Yet the jurisprudential record more pits the First Amendment right of broadcasters against the need of government to "assist parents" as best they can. There has been no direct call from the courts—as there has been by the FCC—to demand "social responsibility" in the form of self-regulation.[11]

Regulation of Language

Serious conceptual difficulties emerge when a governmental entity attempts to define media regulation in terms of language.[12] In areas outside of broadcasting, the academic community has begun to understand that definitions carry with them evaluative interpretations. Thus, attempts to apply exacting definitions to broadcast content considered "indecent" may be doomed.

Obscenity: A Special Case

Any discussion of broadcast indecency must first distinguish it—at least as precisely as possible for legal reasons—from obscenity.

Obscenity has not gained constitutional protection. It has been defined by the U.S. Supreme Court in *Miller v. California* (1973) as that content that appeals to prurient interest based upon contemporary community standards, depicts patently offensive sexual conduct, and lacks serious value.

In the case of broadcasting, however, the waters are muddied by the fact that the obscenity-indecency dichotomy is not followed in the Communications Act of 1934 or a separate criminal law, 18 U.S.C. Sec. 1464, which lumps together obscene, indecent, and profane language. The working definition of broadcast indecency is: *"Language or material that depicts or describes, in terms patently offensive as measured by contemporary standards for the broadcast medium, sexual or excretory activities or organs*." In short, while obscenity must involve sexual conduct, indecency can be merely sexual discussion.

The Legacy of *Pacifica*: Revisiting the Decision

A single complaint apparently prompted action that led to the most significant legal discussion to date on broadcast indecency. By now, most are familiar with the case. A George Carlin monologue—"Filthy Words"—was broadcast on a New York station. The station responded to FCC inquiries by defending Carlin as a "social satirist."[13]

The single broadcast, however, did not lead to direct FCC action; instead, it merely produced an order on file that could be considered if there were future complaints against the station. The FCC found broadcasting unique because of traditional scarcity arguments. Because not everyone who wants a broadcast license may have one, the argument goes, those who do have special responsibilities to serve the "public interest, convenience, and necessity." In recent years, however, economic theorists have argued that all goods are scarce resources—including the paper it requires to publish a newspaper.[14]

Broadcast regulators have also argued that access by children is unsupervised, screening of broadcasts by parents is difficult, and privacy at home justifies control over broadcast indecency. The FCC action against the 2 P.M. *Pacifica* broadcast of the Carlin monologue was not seen by the FCC or the Supreme Court as enforcement of an outright ban but rather as part of a "channeling" policy to move indecent programs to times of day when children were not likely to be in the audience.

It was not clear, however, how the channeling approach could be squared with Section 1464—the criminal statute that seemed to ban broadcast indecency. In part, the conceptual tip-toeing emerged because Section 1464 was not consistent with a separate prohibition against "censorship" of broadcasting found in Section 326 of the Communications Act of 1934.

A divided U.S. Supreme Court in *Pacifica* found: "A requirement that indecent language be avoided will have its primary effect on form, rather than content, of serious communication" and: "There are few, if any, thoughts that cannot be expressed by the use of less offensive language."[15] But a dissenting opinion noted that "taboo surrounding the words" in the Carlin mono-logue was not universal. Justice Brennan rejected the idea that alternative words are always useful: "[I]t is doubtful that the steril-ized message will convey the emotion that is an essential part of so many communications."[16]

Following *Pacifica*, FCC enforcement "was nonexistent" until 1987. Action against individual stations, and a changing political environment, led to movement toward a more clearly defined policy.[17]

Politics and the Twenty-Four-Hour Ban

In late 1988, when the Federal Communications Commission shifted under congressional pressure and issued a twenty-four-hour ban on indecent broadcasts, the FCC made it clear that the new Section 1464 enforcement rule came, "Pursuant to a recent Congressional directive," and that the enforcement was "required by the express language of this new legislation."[18] It has long been understood that communications policy operates within a political context. The rule amounted to a reinterpretation of 18 U.S.C. 1464 that previously had been applied "to prohibit the broadcast of obscene programming during the entire day and indecent pro-gramming only when there was a reasonable risk that children might be in the audience."[19]

FCC Order and the Diaz Dennis Statement

On December 19, 1988, the FCC adopted an order to enforce 18 U.S.C. 1464 "on a 24 hours per day basis." Noting that the agency funding bill—Pub. L. No. 100-459—signed by then President Reagan October 1, 1988, required the FCC to promulgate such an order, the agency fell into "compliance."

The FCC was forced to ignore the precedential value of rules that prohibited "obscene" broadcasts but allowed "indecent" material when the risk was minimal that children were in the audi-ence. Commissioner Patricia Diaz Dennis, in a separate statement, raised the critical issue: "I have serious doubts whether our new rule will pass constitutional muster."[20] She noted that the *Pacifica* ruling was narrow, emphasizing the time of day of a broadcast as a variable.

At the same time that the FCC was faced with the continuing precedent of *Pacifica*, it also dealt with the U.S. Court of Appeals for the District of Columbia: it had ruled that the FCC had not justified even a 10 P.M. to 6 A.M. rule for "channeling" indecent speech. How then, Diaz Dennis wondered, could "an outright ban" be justified?

Dial-A-Porn: *Sable Communications*

For the first time since the late 1970s, we heard from the U.S. Supreme Court on indecency. However, the Court used the telephone case to distinguish broadcasting and hold its *Pacifica* ground. In the case of "dial-a-porn" telephone messages, the Supreme Court in 1989 ruled that indecent but not obscene messages are constitutionally protected.[21] In striking down a congressional ban, the Court refused to define the difference between the two types of speech. Further, the Court balked at the opportunity to directly use the case as a vehicle to rethink the *Pacifica* ruling. But at the same time it constructed language that the lower court could use to strike down the twenty-four-hour ban.

In *Sable Communications v. FCC*, in an opinion delivered by Justice White, the Court treated telephone indecency and broadcast indecency as "distinguishable" because of "an emphatically narrow holding" in *Pacifica*:

> Pacifica is readily distinguishable from this case, most obviously because it did not involve a total ban on broadcasting indecent material. The FCC rule was not "intended to place an absolute prohibition on the broadcast of this type of language, but rather sought to channel it to times of day when children most likely would not be exposed to it."

In *Pacifica*: "The issue of a total ban was not before the court." Still, the precedential value of *Sable* was minimal in the distinct case of broadcast indecency because of the *Pacifica* finding that broadcasting has a "uniquely pervasive" ability to "intrude on the privacy of the home without warning as to program content"—content parents might want to keep out of reach of small children: Placing a telephone call is not the same as turning on a radio and being taken by surprise by an indecent message.

Unlike an unexpected outburst on a radio broadcast, the message received by one who places a call to a dial-a-porn service is not so invasive or surprising that it prevents an unwilling listener from avoiding exposure to it. While the Court did not tamper with *Pacifica*, it surely did provide the lower court with a basis to strike down a twenty-four-hour ban by reiterating in the broadcast context that: "[T]he government may not 'reduce the adult population . . . to . . . only what is fit for children.' "

The Court, then, took a functional approach rather than a content-based approach by accepting the notion that government is free to regulate to "protect" just children: "For all we know from this record, the FCC's technological approach to restricting dial-a-porn messages to adults who seek them would be extremely effective, and only a few of the most enterprising and disobedient young people will manage to secure access to such messages," the Court said. That would be acceptable, in the view of the *Sable* Court, if it were not for the concurrent effect of limiting access to adults: "It is another case of 'burn[ing] up the house to roast the pig.' "

In a concurrence, Justice Scalia agreed with the view "that a wholesale prohibition upon adult access to indecent speech cannot be adopted merely because the FCC's alternate proposal would be circumvented by as few children as the evidence suggests." But he went further in appearing to question the shaky legal distinction between indecency and obscenity:

> But where a reasonable person draws the line in this balancing process—that is, how few children render the risk unacceptable—depends in part upon what mere "indecency" (as opposed to "obscenity") includes. The more narrow the understanding of what is "obscene," and hence the more pornographic what is embraced within the residual category of "indecency," the more reasonable it becomes to insist upon greater assurance of insulation from minors.

The theory advanced here is complex. First, must one view "indecency" as residual or "left over" speech, protected only because it is not obscene? Or does it make just as much sense to view it as that which has not crossed a prohibited government line? Second, must we think of indecency-obscenity as a continuum? If so, what specific elements contribute to an increasing level of prohibitive speech or behavior? Third, on what grounds does an increasing degree of obscene material warrant stiffer safeguards?

The concurrence in part, dissent in part, of Justice Brennan, joined by Justices Marshall and Stevens, spoke more to the issue of definitional problems. Citing his own words in *Paris Adult Theatre I*, Brennan repeated: "[T]he concept of 'obscenity' cannot be defined with sufficient specificity and clarity to provide fair notice to persons who create and distribute sexually oriented materials, to prevent substantial erosion of protected speech as a by-product of the attempt to suppress unprotected speech, and to avoid very costly institutional harms."[22] Here, the functional concern is with a chilling effect on producers of material irrespective of how specific audience members view it as "indecent" or "obscene." The Court, having failed to specify a definition—perhaps because it is so problematic—retreated. But does not "the marketplace" dictate that all producers must face the whims of audience taste? This is so, but the goal of Brennan appears to be to clear the way for the audience to do this, without government regulation exercising a sifting role before hand: "Hence, the Government cannot plausibly claim that its legitimate interest in protecting children warrants this draconian restriction on the First Amendment rights of adults who seek to hear the messages that Sable and others provide."

Treatment of Broadcasting

Even before its decision on the twenty-four-hour ban—the so-called "Helm's amendment" or "congressional mandate"—the issue was a potent one. The previous FCC attempt to move from a midnight "channeling" to a 10 P.M. "safe harbor" approach for indecent broadcasts was not endorsed.

By May 1989, the FCC's Mass Media Bureau had drafted "letters of inquiry," essentially charging stations with violating rules by transmitting indecent broadcasts. With the legal issues unresolved, *Broadcasting* magazine reported that the "FCC believes it can take action against 'daytime' broadcasts only." In October 1989, ninety-five complaints were "disposed of by the Commission," while "notices of apparent liability" were sent to a handful of stations. Stations KFI, Los Angeles, WIOD and WZTA-FM, Miami, and KLUC-FM, Las Vegas, faced action. Four other stations sought additional information. In November, an additional liability letter was sent to WLUP, Chicago.[23]

The 1989 FCC letters to stations came in several waves. There were three letters on August 24, seven letters on October 26, and two follow-up letters—on November 30, 1989 and on January 17, 1990.[24]

The August 24th Letters. Evergreen Media Corp.(WLUP-AM) received complaints about the *Steve and Gary Show* broadcast in Chicago.[25] The licensee was reminded about Section 1464 prohibitions against indecent language and the policy of the FCC to enforce it.[26] The Complaints and Investigations Branch alleged: "[I]t appears that, in context, the material broadcast is 'clear and capable of a specific, sexual meaning' and is patently offensive."[27]

The FCC cited a transcript from a March 30, 1989 broadcast at 5:10 P.M.:

> **Bruce Wolfe:** He's (Bob Costas) trying to defend this Vanessa Williams. I mean she's a, the most embarrassing pictures that you ever saw in your life in Penthouse. I don't know and he's talking about how she actually had more talent than your typical Miss America. I'll say.
> **Steve Dahl:** She was licking that other woman's vagina. I want to tell you pal.

The FCC also cited text from an August 19 broadcast in which a caller talked about "kiddie porn" and joked about a "gay bar." Using the same indecency standards, the FCC sent similar letters to other stations on August 24. KSJO, San Jose's *The Perry Stone Show* was cited for a series of alleged violations during the morning drive shift in 1988. Some of the incidents involved exchanges with callers:

> **Caller:** What do you want?
> **Perry Stone:** I love you. I'd love to, I'd love to lick the matzo balls right off your butt. As a matter of fact, I'd put it right there in the middle.

Stone was also cited for use of unfit song lyrics such as: "My pussy cat was rocking in the rocking chair, rocked so long he lost his hair. . . ."

WFBQ-FM, Indianapolis—licensed by Great American TV in Cincinnati—ran afoul with its *Bob and Tom Show* in September 1987.

> **Elvis:** So you talk to Dick Nixon, man you get him on the phone and Dick suggests maybe getting like a mega-Dick to help

out, but you know, you remember the time the king ate mega-Dick under the table at a Q95 picnic, so that's pretty much out of the question. And then you think about getting Mega-Hodgie, but that's no good because you know, the king was a karate dude and although Mega-Hodgie can take a punch. . . .

The station's parody problems extended to the the language used in fake commercial spots:

Introducing Butch beer, the first beer brewed for women, by women. When you grab a Butch beer, you're taking hold of the Billie Jean King of beers. Fire brewed from the gushing waters of French Lick, Indiana. With Butch beer, you've got a beer that goes down easy. Taste it and you'll know why it's our personal best. . . .

Such parody spots that do not clearly use foul language but depend upon implicit understandings may create the most difficult en-forcement issues for the FCC. The WFBQ-FM case led to a follow-up letter on January 17. At issue was the repetition of already noted material in the previous FCC letter:

The Commission now has information that at least two segments which were cited in the letter of inquiry were repeated on the *Bob and Tom Show*. Specifically, the segments originally broadcast on September 25, and June 29, 1987 . . . were again broadcast between 6:00 A.M. and approximately 10:15 A.M. on August 25, 1989. The Commission also believes that new material broadcast on March 21 and April 13, 1989, by WFBQ(FM) may have violated 18 U.S.C. Section 1464 (transcripts attached).

The charge of new material would be more serious for WFBQ since the licensee could always make the case that the August 24 letter had not been reviewed at the time of the August 25 broadcast. A station already under investigation, would be expected by the FCC to take greater care in the production of new material.

The October 26th Letters. Complaints leveled against Cox Broadcasting, licensee for WIOD, Miami, followed the same gene-ral tone of the earlier complaint letters. That station admitted also airing a "Butch beer commercial" on the *Neil Rogers Show*. The Commission Chief of the Mass Media Bureau, Roy J. Stewart, wrote: "Even in the cases of innuendo, not only was the language understandable and clearly capable of a specific sexual meaning but, because of the context, the sexual import was inescapable." Stewart said reasonableness of context was no defense, and the station could not rely on the safety of lack of action by the FCC:

> [W]e have repeatedly stated that, mindful of the sensitive First Amendment issues involved in indecency rulings, we strive to proceed with caution and careful deliberation, sometimes resulting, unfortunately, in enforcement actions and opinions that reach year-old broadcasts. Whether or not the context of the entire *Neil Rogers Show* dwelt on sexual themes, the songs themselves provide sufficient context to determine their patent offensiveness and can be considered discrete units for purposes of this action. And whether or not the material here at issue is less graphic than that previously found indecent by the Commission, we do not accept constraints on our discretion to pursue violations less egregious than others have been.

The FCC rejected the licensee's arguments that the broadcasts of the songs "Jet Boy, Jet Girl," "Penis Envy," "Candy Wrapper," and "Walk with an Erection" fit with "contemporary community standards," were "acceptable slang," and were supported by the evidence of audience ratings.

Broadcast indecency is a violation of *federal law* and its popularity in any particular community does not change that fact. But more importantly, the focus of our indecency standards must be on the *risk to children* in the audience. WIOD, Miami was fined the maximum $10,000 penalty.

Station WZTA-FM, Miami—operated by Guy Gannett Publishing—also received on October 26 letter based on its broadcast of the "Penis Envy Song."

Station KLUC-FM—licensed to Nationwide Communications—was notified of a complaint about the playing of the Prince/Warner Brothers song "Erotic City." That song carries the lyrics: "If we cannot make babies maybe we can make some time . . . fuck so pretty you and me, erotic city come alive." The case differed from the previously described actions in that: (1) it involved a commercially distributed song, and (2) it involved usage of one of the "seven dirty words"—namely "fuck."

WXRK-FM, New York also received an October 26 letter from the FCC. The complaint focused on broadcasts by Howard Stern on Infinity stations WXRK-FM and WYSP-FM, Philadelphia, and WJFK-FM, Washington, D.C., in 1988. In part, one of the broadcasts aired contained this sexual material:

I want to rub my ear and have this girl go wild for me. . . . When we come back from commercial, we have a young man who wants to play the piano with his, uh, wiener. . . . Howard, I'd better go into the other room and, uh, get it ready. I'll come in swinging it. It's bigger than yours. I've got a rubber. Don't worry about it. And I've got a second rubber for encore. He's going to wear a contraceptive. I do safe organisms . . . orgasims. I'm going to play Casio. . . . I believe we hit two keys at the same time. You'd better give me the next segment, though. I'm going to get it going. O.K.? Go in the other room and do whatever you have to do to play it.

The complaint against Diamond Broadcasting's Paris, Arkansas, stations KCCL-AM/FM was different from the other October 26 complaint letters in that it involved what was apparently unplanned programming. According to the letter a phone conversation between general manager Gene Williams and his son was broadcast live at 7:00 P.M. on January 7, 1988:

> Mr. William's son allegedly stated, "To all you listeners in Paris, Arkansas, don't bend over in front of my dad. Gene Williams will fuck you in the ass," at which point the phone was unplugged by another station employee. Mr. Williams apparently phoned his son back and over the air stated, "You fucking asshole, I told you we were on the air," to which his son replied, "I don't give a damn." This conversation contained material which, when broadcast, may have violated 18 U.S.C. Section 1464.

KFI-AM, a Los Angeles station, also received an October 26, 1989 letter about material broadcast during Tom Leykis's 11-3 midday shift on November 12, 1988, and January 6, 1989.

> **Female Caller:** I'm a breeder-exhibitor of very large dogs and I have five beautiful dogs and my male, sometimes when I groom him, I'll masturbate him and he's just so big and sweet and he really likes it. It's really weird, I know. I could never tell anybody.
> **Tom Leykis:** Oh boy, now, now wait a minute. You masturbate just one dog or do you do this with more than one?
> **Female Caller:** Just the male. I have a big champion male and he's just beautiful and you know, I'll pet him and I don't know, he really likes it.
> **Tom Leykis:** Now, do you enjoy masturbating your dog?
> **Female Caller:** Well, I do because it's, he likes it, I mean, you should see the expression on his face.

Yet another of the October 26 letters went to WWWE-AM, Cleveland about the *Gary Dee Show* and a 2 P.M. broadcast on June 15, 1989:

> **Caller:** My comment is Lou told me that Chris James gives fellatio okay. Now my dream, ya still there, Gary?
> **Gary Dee:** Yeah.
> **Caller:** Now, my dream is to get in Chris James' pants.
> **Gary Dee:** Well, why don't you be a man and go down there and say that to her face. You know, you are just like a little kid masturbating in public. You've got no class, you've got no taste.

The final October 26 letter went to KSD-FM, St. Louis, and licensee Pacific and Southern. Excerpts from an October 1987 *Playboy* were said to have been aired at 6:50 A.M. on September 27, 1987.

The November 30 WLUP(AM) Letter. After WLUP(AM) responded October 10, 1989 to the FCC that the *Steve and Gary Show* was not indecent in the view of Evergreen Media, the FCC issued a notice of forfeiture.[28] The station had claimed that the show was "extemporaneous, open forum, frank, live, and spontaneous, often humorous, with over half the dialogue supplied by listeners subjected to WLUP's reasonable seven-second screenings. . . ."[29] The station further argued that the program is aimed at adults, and there is no evidence that children listen. The station argued FCC action would create a chilling effect, and that listeners in Chicago did not find the show patently offensive. The FCC responded that talk about homosexual activity, child pornography, and oral-genital contact are indecent:

> We believe that all the subject broadcasts fit squarely within our definition of indecency quoted above. The March 30, 1989, "straightforward description" of sexual activities that appeared in the magazine (the sale of which could very well have been restricted as to minors) was delivered by the talk show host in explicit, graphic and vulgar language, at a time of day when unsupervised children were likely to have been in the listening audience. We do not accept your argument that the broadcast's asserted value as political and social commentary should shield it from normal Commission scrutiny or place it in a special category less vulnerable to Commission sanction. We also believe that the innuendo of that broadcast, as well as the innuendo and double entendre in the two others, was understandable and clearly capable of a specific sexual meaning, the import of which was inescapable.[30]

WLUP-AM, thus, faced a $6,000 fine. But the FCC has yet to clarify its degrees of difference between a $2,000, $6,000, or $10,000 fine in indecency cases.

Future Proposals: Is There a Way Out?

The industry has been highly critical of the FCC's latest attempts to restrict indecent language, but it has been unwilling to advance useful alternatives. The trade magazine *Broadcasting* (now called *Broadcasting & Cable*) found that the FCC was waving a red flag in a "prurient pursuit" when it should have been dealing with more serious industry issues.[31] The NAB (National Association of Broadcasters) argued a flat ban, as proposed, was unconstitutional.[32] And the RTNDA (Radio-Television News Directors Association) viewed the proposed ban as a First Amendment issue by preventing "the occasional use of news subjects' 'dirty' language in late evening newscasts, documentaries, interviews, and talk and 'magazine' shows."[33]

Meanwhile, the U.S. Court of Appeals, District of Columbia Circuit, seemed to find the proposed ban as in conflict with that court's earlier ruling—an opinion supporting time channeling rather than a content ban.[34] And, the FCC was left struggling with a lack of evidence to support effectiveness of either a channeling approach or total ban.[35] So, is there a way out? On the face of it, the FCC's historic tradition is against strict enforcement of Section 1464; in only a handful of cases has the FCC imposed liability, and usually in the form of small fines. The record of inaction was dramatic.

The FCC Record of Inaction

In one of the earliest cases, the FCC found that WUHY-FM had erred in the broadcast of a taped interview containing words such as "shit" and "fuck," but—despite evidence that teens were in the audience—the fine was just $100. In a dissent by Commissioner Kenneth A. Cox to even this action, the fear of broadcaster timidity in reaction to the FCC action was expressed:

> WUHY received no complaints about the broadcast here in question, nor did the Commission. However, we had received earlier com-

plaints about the 10 to 11 P.M. time period and were monitoring the station on the night of January 4, 1970. So far as I can tell you, my colleagues are the only people who have encountered this program who are greatly disturbed by it.[36]

Under the rule, the FCC argued for a $2,000 fine in cases where violations were repeated or willful. But in the 1973 WGLD-FM "topless" radio case, that higher level of fine was justified only because the interviewer could have moderated "his handling of the subject matter so as to conform to the basic statutory standards."[37] It was not made clear how discussion of oral sex as a method to keep one's sex life alive could have been moderated, but the dissent of Commissioner Nicholas Johnson called for broadcaster discretion. Johnson identified the inherent problem of treating indecent language as obscenity: "The majority admits that 'indecent' expression is something less than obscenity, yet the majority nevertheless asserts that it may outlaw indecent expression."[38] Johnson said the definition for obscenity is vague, and "if obscenity is so vaguely defined, then the 'indecency' variant promulgated by the majority is a hopeless blur." Johnson refused to participate in a closed-door review of monitoring tapes, and he said the FCC in doing so was acting as a "Big Brother" programming review board "allegedly capable of deciding what is and is not good for the American people to see and hear."

When Sonderling Broadcasting paid its $2,000 fine in 1973, the station denied liability.[39] The FCC stood by its view that by replacing a general indecency definition of "utterly without redeeming social value," to the more narrow Miller language, the FCC was on safe legal footing. The federal court in the *Illinois Citizens Committee* case upheld the FCC fine under the view that the program was obscene not indecent.[40]

In the *Yale Broadcasting* case the FCC never prohibited drug-oriented language; the FCC only required stations to monitor their broadcasts.[41]

In the FCC's 1975 report, the FCC continued to tread carefully. It was only Commissioner Charlotte T. Reid, in a concurrence, who argued the position that is alive today:

> While I am particularly shocked that such language was broadcast at a time when children could be expected to be in the audience, I feel constrained to point out that I believe this language to be totally inappropriate for broadcast at any time. In this sense, I think that the Commission's standards do not go far enough. To me, the language

used in this case has absolutely no place on the air whether it be 2:00 P.M. or 2:00 A.M.[42]

Such a high point in support of regulation might be seen in the action against University of Pennsylvania station WXPN-FM if it were not for the ultimate finding that it was licensee failure to supervise, and not the language itself, that was of the greatest concern. The 1975 broadcasts noted by the FCC used terms such as "fucker," "pussy," and "ass," but the FCC seemed most struck by a telephone caller who said he was three years old. Even here, however, the initial $2,000 fine was for indecent and obscene broadcasts.[43]

By the time of the *Infinity* case in 1987, the FCC was of the view that "indecent" language is broader than the Carlin seven dirty words, including talk of the penis and animal sex.[44] It is clear, however, that much of what the FCC lumps with indecency could rightly fall outside the protected umbrella of indecency by calling it obscenity. In this way, the print- broadcast dichotomy would be removed. Indecent language would be allowed on the broadcast airwaves at all times of day, but obscene broadcasts would be strictly prohibited. The only legal stumbling block is the language of Section 1464, which has never been directly challenged. Because it is unlikely—given current political sentiment—that the Congress would repeal the obscene, indecent, and profane lumping language, a direct challenge in the courts is needed. The U.S. Court of Appeals has already used language in *ACT I* (the first of four cases titled *Action for Children's Television*, after a Washington-based interest group of the same name) that supports the view against a total ban.[45] Pursuit of this action, then, seemed to move toward the day when the U.S. Supreme Court would ultimately revisit *Pacifica*.[46]

The record has shown great hesitancy on the part of the FCC to take strong action against stations for use of language, but where this has been the case, obscenity—not indecency—has been at issue. This is important for broadcast managers to understand. In typical broadcast situtations, managers operating in good faith should not fear FCC sanctions.

It does not appear that new meaningful principles emerged following the *Pacifica* decision that serve to guide the FCC in its indecency actions. The FCC's late 1980s cases paralleled the cases of the early 1970s.[47]

A coherent policy would leave indecent language as permissible on the broadcast airwaves, but it would continue the ban on obscenity. This would serve to square the content regulation with regulation of printed materials and nonbroadcast electronic media such as cable; and it would eliminate the need to debate time-of-day issues.

No known societal values can be shown that support the need to keep children away from indecent language. It is a different situation from obscenity where behavioral research might suggest danger of negative modeling effects.[48] No such evidence exists in the case of indecent language. In fact, language on the broadcast airwaves, the juke box, or the radio is no doubt "cleaner" than that found in some homes.

In conclusion, the statutory authority of 18 U.S.C. 1464 must be challenged because it is in conflict with the view that broadcasters have first amendment rights of expression. And beyond the strict legal problem, the FCC has shown no interest in employing systematic methods of content analysis that might serve to define what it is about the content that is objectionable. The FCC, to the contrary, historically relies on the good faith judgment of licensees to police their own airwaves. And where complaints surface, the FCC is generally reluctant to step in and enforce with a heavy hand or move to revoke a valuable license. Thus, the FCC has been less than completely honest for over twenty years, purporting to "regulate" indecent language, but mostly serving as a pointless lip service to a Congress it reports to and looks to for readings of the political climate. Some of this may serve the interests of politicians, regulators, and even the industry, but it probably does not serve the interests of the majority of audience members.

Another Look at *Pacifica*

Two decades after the Supreme Court's decision in the *Pacifica* case, the ruling both continues to stand as defining our law of broadcast indecency at the same time as it continues to confuse our law of broadcast indecency. Most, over the years, have become aware of the facts in the case:

At two o'clock in the afternoon on Tuesday, October 30, 1973, Pacifica Foundation's New York radio station, WBAI-FM, broadcast a twelve minute satirical monologue by comedian George Carlin entitled "Filthy Words." WBAI broadcast the monologue, which had been recorded before a live audience in a California theater, as part of a general discussion of contemporary society's attitudes toward language. Prior to the broadcast, listeners were informed that it included sensitive language that some might find offensive and that those who might be offended should change the station and return to WBAI in fifteen minutes. George Carlin began by describing the monologue as being about "the words you couldn't say on the public, ah, airwaves, um, the ones you definitely wouldn't say, ever." He then proceeded to list and expound upon the "seven dirty words."[49]

A motorist driving with his "young son" complained in writing to the FCC. A little-known fact about the complaint—the only one received about the broadcast—was the political nature of it:

> The complaint was made by a Florida resident who lived outside the range of the station's signal and who was a member of the national planning board of Morality in Media. His "young son" who was with him in the car when he heard the monologue was fifteen years old.[50]

The station's response to the complaint can be boiled down to two defenses: (1) listeners had been warned in advance of the broadcast, and (2) the broadcast was "satire" about social attitudes on language.

Both the United States Court of Appeals, District of Columbia Circuit, which ruled in favor of the station, and the Supreme Court, which reversed that decision and supported the FCC action, reproduced the text of the Carlin monologue as an appendix to the opinions. The language in question is reproduced below:

The following is a verbatim transcript of "Filthy Words" (Cut 5, Side 2), from the record album **George Carlin, Occupation: Foole** *(Little David Records, LD 1005).*

"Aruba-du, ruba-tu, ruba-tu.

I was thinking about the curse words and the swear words, the cuss words and the words that you can't say, that you're not supposed to say all the time, 'cause words or people into words want to hear

---

your words. Some guys like to record your words and sell them back to you if they can, (laughter) listen in on the telephone, write down what words you say. A guy who used to be in Washington knew that his phone was tapped, used to answer, Fuck Hoover, yes, go ahead. (laughter) Okay. I was thinking one night about the words you couldn't say on the public, ah, airwaves, um, the ones you definitely couldn't say, ever, 'cause I heard a lady say bitch one night on television, and it was cool like she was talking about, you know, ah, well, the bitch is the first one to notice that in the litter Johnie right (murmur) Right. And, uh, bastard you can say, and hell and damn so I have to figure out which ones you couldn't and ever and it came down to seven but the list is open to amendment and in fact has been changed, uh, by now, ha, a lot of people pointed things out to me, and I noticed some myself. The original seven words were, shit, piss, fuck, cunt, cocksucker, motherfucker, and tits. Those are the ones that will curve your spine, grow hair on your hands, and (laughter) maybe, even bring us, God help us, peace without honor (laughter) um, and a bourbon. (laughter) And now the first thing that we noticed was that the word fuck was really repeated in there because the word motherfucker is a compound word and it's another form of the word fuck. (laughter) You want to be a purist it doesn't really—it can't be on the list of basic words. Also, cocksucker is a compound word and neither half of that is really dirty. The word—the half sucker that's merely suggestive (laughter) and the word cock is a halfway dirty word, 50 percent dirty—dirty half the time, depending on what you mean by it. (laughter) Uh, remember when you first heard it, like in sixth grade, you used to giggle. And the cock crowed three times, heh (laughter) the cock—three times. It's in the Bible, cock is in the Bible. (laughter) And the first time you heard about a cockfight, remember—What? Huh? Naw. It ain't that, are you stupid? Man, (laughter, clapping) it's chickens, you know. (laughter) Then you have the four letter words from the old Anglo-Saxon fame. Uh, shit and fuck. The word shit, uh, is an interesting kind of word in that the middle class has never really accepted it and approved it. They use it like crazy, but it's not really okay. It's still a rude, dirty, old kind of gushy word. (laughter) They don't like that, but they say it, like, they say it like, a lady now in a middle-class home, you'll hear most of the time she says it as an expletive, you know, it's out of her mouth before she knows. She says, Oh shit oh shit, (laughter) oh shit. If she drops something, Oh, the shit hurt the broccoli. Shit. Thank you. (footsteps fading away) (papers ruffling) Read it! (from audience) Shit! (laughter) I won the Grammy, man, for the comedy album. Isn't that groovy? (clapping, whistling) (murmur) That's true. Thank you. Thank you man. Yeah. (murmur) (continuous clapping) Thank you man. Thank you. Thank you very much, man. Thank—no, (end of continuous clapping) for that and for the Grammy, man, cause (laughter) that's based on people liking it man, yeh, that's ah, that's okay man. (laughter) Let's let that go, man. I got my Grammy, I can let my hair hang down now, shit. (laughter) Ha! So! Now the word shit is okay for the man. At work you can say it like crazy. Mostly figuratively, Get that shit out of here, will ya? I don't want to see that shit anymore. I can't cut that shit, buddy. I've had that shit up to here. I think you're full of shit myself. (laughter) He don't know shit from Shinola (laughter) you know that? (laughter) Always wondered how the Shinola people felt about that. (laughter) Hi, I'm the new man from Shinola. (laughter) Hi, how are ya? Nice to see ya. (laughter) How are ya? Boy, I don't know whether to shit or wind my watch. (laughter) Guess, I'll shit on my watch. (laughter) Oh, the shit is going to hit de fan. (laughter) Built like a brick shit-house. (laughter) Up, he's up shit's creek. (laughter) He's had it. (laughter) He hit me, I'm sorry. (laughter) Hot shit, holy shit, tough shit, eat shit, (laughter) shit-eating grin. Uh, whoever thought of that was ill. (murmur laughter) He had a shit-eating grin! He had a what? (laughter) Shit on a stick. (laughter) Shit in a handbag. I always liked that. He ain't worth shit in a handbag. (laughter) Shitty. He acted real shitty. (laughter) You know what I mean? (laughter) I got the money back, but a real shitty attitude. Heh, he had a shit-fit. (laughter) Wow! Shit-fit. Whew! Glad I wasn't there. (murmur, laughter) All the animals—bull shit, horse shit, cow shit, rat shit, bat shit. (laughter) First time I heard bat shit, I really came apart. A guy in a Oklahoma, Boggs, said it, man. Aw! Bat shit. (laughter) Vera reminded me of that last night, ah (murmur). Snake shit, slicker than owl shit. (laughter) Get your shit together. Shit or get off the pot. (laughter) I got a shit-load full of them. (laughter) I got a shit-pot full, all right. Shithead, shitheel, shit in your heart, shit for brains, (laughter) shit-face, heh. (laughter) I always try to think how that could have originated; the first guy that said that. Somebody got drunk and fell in some shit, you know. (laughter) Hey, I'm shit-face. (laughter) Shit-face, today. (laughter) Anyway, enough of that shit. (laughter) The big one, the word fuck that's the one that hangs them up the most. 'Cause in a lot of cases that's the very act that hangs them up the most. So, it's natural that the word would, uh, have the same effect. It's a great word, fuck, nice word, easy word, cute word, kind of. Easy word to say. One syllable, short u. (laughter) Fuck. (Murmur) You know, it's easy. Starts with a nice soft sound fuh ends with a kuh. Right? (laughter) A little something for everyone. Fuck. (laughter) Good word. Kind of a proud word, too. Who are you? I am FUCK. (laughter) FUCK OF THE MOUNTAIN. (laughter) Tune in again next week to FUCK OF THE MOUNTAIN. (laughter) It's an interesting word too, 'cause it's got a double kind of a life— personality—dual, you know, whatever the right phrase is. It leads a double life, the word fuck. First of all, it means, sometimes, most of the time, fuck. What does it mean? It means to make love. Right? We're going to make love, yeh, we're going to fuck, yeh, we're going to fuck, yeh, we're going to make love, (laughter) we're really going to fuck, yeh, we're going to make love. Right? And it also means the beginning of life, it's the act that begins life, so there's the word hanging around with words like love, and life, and yet on the other hand, it's also a word that we really use to hurt each other with, man. It's a heavy. It's one that you save toward the end of the argu- ment. (laughter) Right? (laughter) You finally can't make out. Oh, fuck you man. I said, fuck you. (laughter, murmur) Stupid fuck. (laughter) Fuck you and everybody that looks like you, (laughter) man. It would be nice to change the movies that we already have and substitute the word fuck for the word kill, wherever we could, and some of those movie cliches would change a little bit. Mad fuck- ers still on the loose. Stop me before I fuck again. Fuck the ump, fuck the ump, fuck the ump, fuck the ump, fuck the ump. Easy on the clutch Bill, you'll fuck that engine again. (laughter) The other shit one was, I don't give a shit. Like it's worth something, you know? (laughter) I don't give a shit. Hey, well, I don't take no shit, (laugh- ter) you know what I mean? You know why I don't take no shit? (laughter) Cause I don't give a shit. (laughter) If I give a shit, I would have to pack shit. (laughter) But I don't pack no shit cause I don't give a shit. (laughter) You wouldn't shit me, would you? (laughter) That's a joke when you're a kid with a worm looking out the bird's ass. You wouldn't shit me, would you? (laughter) It's an eight-year- old joke but a good one. (laughter) The additions to the list, I found three more words that had to be put on the list of words you could never say on television, and they were fart, turd, and twat, those three. (laughter) Fart, we talked about, it's harmless. It's like tits, it's a cutie word, no problem. Turd, you can't say but who wants to, you know? (laughter) The subject never comes up on the panel so I'm not worried about that one. Now the word twat is an interesting word. Twat! Yeh, right in the twat. (laughter) Twat is an interesting word because it's the only one I know of, the only slang word apply- ing to the, a part of the sexual anatomy that doesn't have another meaning to it. Like, ah, snatch, box, and pussy all have other meanings, man. Even in a Walt Disney movie, you can say, We're going to snatch that pussy and put him in a box and bring him on the airplane. (murmur, laughter) Everybody loves it. The twat stands alone, man, as it should. And two-way words. As, ass is okay providing you're riding into town on a religious feast day. (laughter) You can't say, up your ass. (laughter) You can say, stuff it! (murmur) There are certain things you can say its weird but you can just come so close. Before I cut, I, uh, want to, ah, thank you for listening to my words, man, fellow, uh, space travelers. Thank you man for tonight, and thank you also. (clapping, whistling)"

The sharply divided Supreme Court ruling was a reversal of what the Court of Appeals had found. Key portions of the appellate case are reproduced on the following pages:

Pacifica Foundation, Petitioner v. Federal Communications Commission and United States of America, Respondents

Pacifica Found. v. FCC No. 75-1391 United States Court of Appeals for the District of Columbia Circuit

556 F.2d 9; 40 Rad. Reg. 2d (P & F) 99; 2 Media L. Rep. 1465
March 30, 1976, Argued
March 16, 1977, Decided
TAMM, Circuit Judge:

This appeal by Pacifica Foundation (*Pacifica*) challenges a Federal Communications Commission (FCC or Commission) ruling which purports to ban prospectively the broadcast, whenever children are in the audience, of language which depicts sexual or excretory activities and organs, specifically seven patently offensive words. Without deciding the perplexing question of whether the FCC, because of the unique characteristics of radio and television, may prohibit non-obscene speech or speech that would otherwise be constitutionally protected, we find that the challenged ruling is overbroad and carries the FCC beyond protection of the public interest into the forbidden realm of censorship. For the reasons which follow, we reverse the Commission's order.

I. Factual Background

On the afternoon of October 30, 1973, Station WBAI, New York, New York (which is licensed to Pacifica), was conducting a general discussion of contemporary society's attitude toward language as part of its regular programming.

On December 3, 1973, the Commission received a complaint from a man in New York stating that, while driving in his car with his young son, he had heard the WBAI broadcast of the Carlin monologue. This was the only complaint lodged with either the FCC or WBAI concerning the Carlin broadcast. The Commission determined that clarification of its definition of the term "indecent" was in order. As a result, in *Pacifica Foundation*, 56 F.C.C.2d 94 (1975) (hereinafter *Order*), the Commission defined as indecent, language that describes, in terms patently offensive as measured by contemporary community standards for the broadcast medium, sexual or excretory activities and organs, at times of the day when there is a reasonable risk that children may be in the audience. The Commission found that the seven four-letter words contained in the Carlin monologue depicted sexual or excretory organs and activities in a patently offensive manner, judged by contemporary community standards for the broadcast medium, and accordingly, were indecent.

In concurring statements, Commissioners Reid and Quello felt the Order did not go far enough. Commissioner Reid believed indecent language was inappropriate for broadcast at any time. Commissioner Quello was in agreement, commenting that "garbage is garbage" and it should all be prohibited from the airwaves. Id. at 102, 103.

Pacifica argues that the Carlin monologue is not obscene because it does not appeal to any prurient interest and because it has literary and political value. Therefore, Pacifica argues it is entitled to constitutional protection in light of *Miller*. . . .

One week prior to oral argument in this case the FCC released a memorandum and order seeking to clarify its earlier Order. The order of clarification was in response to a petition filed by the Radio Television News Directors Association. In the clarification order, the Commission declared that it never intended to place an absolute prohibition on the broadcast of indecent language but only sought to channel it to times of the day when children would least likely be exposed to it.

The clarifying order, in attempting to narrow the scope of the original Order, ruled that indecent language could be broadcast in a news or public affairs program or otherwise if it was aired at a time when the number of children in the audience was reduced to a minimum, if sufficient warning were given to unconsenting adults, and if the language in context had serious literary, artistic, political or scientific value.

The Commission determined that it would be inequitable to hold a licensee responsible for indecent language broadcast during live coverage of a newsmaking event. The Commission thought it better to trust the licensee to exercise judgment, responsibility and sensitivity to the needs, interest, and tastes of the community.

II. Resolution
(this is the court's language in resolving the legal issues in the case)

Despite the Commission's professed intentions, the direct effect of its Order is to inhibit the free and robust exchange of ideas on a wide range of issues and subjects by means of radio and television communications. In promulgating the Order the Commission has ignored both the statute which forbids it to censor radio communications and its own previous decisions and orders which leave the question of programming content to the discretion of the licensee. The Commission claims that its Order does not censor indecent language but rather channels it to certain times of the day. In fact the Order is censorship, regardless of what the Commission chooses to call it. The intent of the Commission is clear. It is to keep language that describes sexual or excretory organs and activities from the airwaves when there is a reasonable risk that children may be in the audience. The Commission expressly states that this language has "no place on radio" and that when children are in the audience a claim that it has literary, artistic, political or scientific value will not redeem it. . . .

As the study cited by the amicus curiae . . . illustrates, large numbers of children are in the broadcast audience until 1:30 A.M. The number of children watching television does not fall below one million until 1:00 A.M. As long as such large numbers of children are in the audience the seven words noted in the Order may not be broadcast. Whether the broadcast containing such words may have serious artistic, literary, political or scientific value has no bearing on the prohibitive effect of the Order. The Commission's action proscribes the uncensored broadcast of many of the great works of literature including Shakespearean plays and contemporary plays which have won critical acclaim, the works of renowned classical and contemporary poets and writers, and passages from the Bible.

The importance of independent judgment by local licensees has been affirmed again and again by the FCC and the courts. Perhaps the most important ruling for our purpose is the Commission's clarification memorandum regarding the original Order. There the Commission recognized that in some cases, public events likely to produce offensive speech are covered live, and there is no opportunity for journalistic editing. Under these circumstances we believe that it would be inequitable for us to hold a licensee responsible for indecent language.

Unquestionably the Commission's Order also raises First Amendment considerations. The Commission recognized that Congress had prohibited it from engaging in censorship or interfering "with the right of free speech by means of radio communication." In the Order, the Commission contends that because of its unique qualities the broadcast medium is not subject to the same constitutional standards that may be applied to other less intrusive forms of expression.

As defined by Congress, and refined by the FCC and the courts, public interest has always been understood to require licensees to offer some balance in their program format. . . . Obviously balanced programming requires more than just programs suitable for children. Speech cannot be stifled by the government merely because it would draw an adverse reaction from the majority of the people.

The Commission assumes that absent FCC action, filth will flood the airwaves. Thus the Commission argues that the alternative of turning the dial will not aid the sensitive person in his efforts to avoid filthy language.

The Order provides no empirical data to substantiate this assumption. Moreover, the assumption ignores the forces of economics and of ratings on the substance of programming. Licensees are businesses and depend on advertising revenues for survival. The corporate profit motive and the connection between advertising revenue and audience size suggest that the dike will hold as long as the community remains actually offended by what it sees or hears. Commentators and commissioners alike have noted that broadcast media require majorities, or at least sizable pluralities, to pay the bills. If they are correct, and if the Commission truly seeks only to enforce community standards, the market should limit the filth accordingly.

Conclusion
(of this court)
   As we find that the Commission's Order is in violation of its duty to avoid censorship of radio communications under *47 U.S.C. @ 326* and that even assuming, arguendo, that the Commission may regulate non-obscene speech, nevertheless its Order is over-broad and vague, therefore we must reverse the Order. We should continue to trust the licensee to exercise judgment, responsibility, and sensitivity to the community's needs, interests and tastes. To whatever extent we err, or the Commission errs in balancing its duties, it must be in favor of preserving the values of free expression and freedom from governmental interference in matters of taste.
   So ordered.

(Concurring opinion filed by Chief Judge Bazelon.)

Conclusion
The impact of television and radio has grown at an astonishing rate, and broadcasting promises to become by far the most influential medium of communications in our society. As its power continues to grow, preservation of free speech will hinge largely on zealously protecting broadcasting from censorship. As Chief Justice Warren once observed, the impact of a particular medium constitutes no basis for subjecting that medium to greater suppression: This is the traditional argument made in the censor's behalf; this is the argument advanced against newspapers at the time of the invention of the printing press. The argument was ultimately rejected in England, and has consistently been held to be contrary

to our Constitution. No compelling reason has been predicated for accepting the contention now.

(Dissenting opinion filed by Circuit Judge Leventhal.)

Applying these considerations to the language used in the monologue broadcast by Pacifica's station WBAI, in New York, the Commission concludes that words such as "fuck," "shit," "piss," "motherfucker," "cocksucker," "cunt" and "tit" depict sexual and excretory activities and organs in a manner patently offensive by contemporary community standards for the broadcast medium and are accordingly "indecent" when broadcast on radio or television. These words were broadcast at a time when children were undoubtedly in the audience (i.e., in the early afternoon).

Moreover, the pre-recorded language with the words repeated over and over was deliberately broadcast. We therefore hold that the language as broadcast was indecent and prohibited by *18 U.S.C. 1464*. Accordingly, the licensee of WBAI-FM could have been the subject of administrative sanctions pursuant to the Communications Act of 1934, as amended. No sanctions will be imposed in connection with this controversy, which has been utilized to clarify the applicable standards. However, this order will be associated with the station's license file, and in the event that subsequent complaints are received, the Commission will then decide whether it should utilize any of the available sanctions it has been granted by Congress. . . . There are several reasons why we are issuing a declaratory order instead of a notice of apparent liability as we did in *WUHY-FM* and *Sonderling*. A declaratory order is a flexible procedural device admirably suited to terminate the present controversy between a listener and the station, and to clarify the standards which the Commission utilizes to judge "indecent language."

Commissioners Reid and Quello stated that in their view the declaratory order should have gone further and prohibited such language at any time of the day or night. That was a minority expression. Chairman Wiley concurred only in the result. Commissioner Robinson, joined by Commissioner Hooks, concurred on the ground of time limitation, as a reasonable measure "to insist that programming of a kind whose broadcast to children would be thought inappropriate be confined to hours of the evening in which children would not ordinarily be exposed to the material—or at least not without the supervision of a parent."

IV. Conclusion (of this opinion)
On the premise advanced by Justice Holmes that "all rights tend to declare themselves absolute to their logical extreme," *Hudson Water Co. v. McCarter* . . . (1908), there is no logical ground for compromise between the right of free speech and the right to have public utterance limited to some outside boundary of decorum. But while the conflicting claims of liberty and propriety cannot be reconciled, they can be made to coexist by tour de force. This agency, in my view, has the power to compel that co-existence in the limited scale we undertake today. I assent to it because I recognize that the only possible way to take a mediate position on issues like obscenity or indecency is to avoid dogmatism and its meretricious handmaiden, the Ringing Phrase, and to split the difference, as sensibly as can be, between the contending ideas.

The majority opinions seem to consider "indecent" as a novel concept in the law, which should in their view not be extended beyond control of the "obscene." They wholly fail to take account of one aspect of *Miller*, which has not been much analyzed but which seems to me to have been deliberate and significant. The pre-*Miller* rulings had always defined "obscene" in terms of what appeals to the lewd and prurient interest, *see e.g. Roth v. United States* . . ., the concept that had previously been defined. But *Miller* expanded on this—to include "patently offensive representations or descriptions of . . . excretory functions."

A concept like "indecent" is not verifiable as a concept of hard science. Its acceptance by and application by the FCC does not necessarily reflect, or depend upon, a determination by the FCC that this material would be dangerous to the children. What it reflects is a determination concerning a broad consensus of society, the view that the great bulk of families would consider it potentially dangerous to their children, and the further view that in our society, with the family as its base block, it is the family that should have the means to make that choice. With the pervasiveness of TV-radio and its reach into the home the choice made by broadcasters precludes an effective choice by the family.

Because of the unique interest in home life, especially strong in homes where children are being raised, it is bootless to quote from cases that reflect a more permissive attitude to speech in public streets and places, without attention to the difference.

A crucial reality, dominating the case at hand, is the widespread access of radio to children. Radio is relatively inexpensive in initial capital cost, and virtually a free good in terms of operating expense. Widespread freedom of selection of programs by children is not only a condition, it is often a necessity. Today, a majority of families with school-age children have working mothers, and one out of five children in the United States live with only one parent, so that many children are at home unsupervised during the day. In this totality of conditions, one cannot wave away the radio-TV problem on the ground that the (mature) person can readily switch the channel.

The abhorrence of Censorship is a vital part of our society. But there is a distinction between the all-out prohibition of a censor, and regulation of time and place of speaking out, which still leaves access to a substantial part of the mature audience. What is entitled to First Amendment protection is not necessarily entitled to First Amendment protection in all places.

Smut may drive itself from the market, and confound Gresham, so Judge Tamm suggests. Judges cannot, however, premise that there is not really a market that will endure. In any event, there is a problem of the transition period. Even the most earnest advocates of freedom accept the role of government in protecting those who lack capacity.

What we have before us is the Federal Communications Commission order declaring the invalidity of particular language "as broadcast." That carries with it the limitations of time and deliberate repetition identified by the FCC.

The limitation of time is the afternoon. I am aware that the FCC's only indication of acceptability for the broadcast referred to the late hours of the evening. But the issue of what might be broadcastable in the early evening is not before us, and raises different considerations. That would be a time when there were large num-

bers of children in home audiences generally, but the issue could be
raised that for homes where parents really care about such matters
there would be at least one parent in a position to monitor the material heard and seen. A ruling expanding the zone of the broadcastable to adult levels might apply when the time of broadcast is
such that the great preponderance of children are subject to parental
control.

The *Pacifica* Ruling

The majority on the Supreme Court supported the FCC position
that the agency had a legal right to control indecent speech on the
public airwaves. It had been previously held in the landmark
National Broadcasting Co. v. United States case that the licensing system of the Communications Act of 1934 was legal because of
"scarcity"—the idea that not everybody who wanted a broadcast
license could have one.[51] The congressional language of the act had
created a vague standard ("public interest, convenience, or necessity")—one borrowed from earlier railroad regulation.

The Act itself establishes that the FCC's powers are not limited
to the engineering and technical aspects of regulation of radio communication. Yet we are asked to regard the FCC as a kind of traffic
officer, policing the wavelengths to prevent stations from interfering with each other. But the Act does not restrict the FCC merely to
the supervision of the traffic. (The *NBC* case upheld the FCC's right
to limit network "chain broadcasting" in a 1940s ruling that
granted expansive FCC powers). It puts upon the FCC the burden
of determining the composition of that traffic.[52]

Our historical understanding of the "broad" powers of the FCC
was further elaborated (as "expansive") by the Supreme Court in
the *Red Lion Broadcasting Co. v. FCC* (an opinion that upheld the
Fairness Doctrine in the 1960s) case: "[T]he people as a whole retain
their interest in free speech by radio and their collective right to
have the medium function consistently with the ends and purposes
of the First Amendment. *It is the right of the viewers and listeners, not
the right of broadcasters, which is paramount. . . ."*[53] (emphasis added)

In this context, the majority position was delivered to broadcasters in an opinion written by Justice Stevens:

The Commission characterized the language used in the Carlin monologue as "patently offensive," though not necessarily obscene, and expressed the opinion that it should be regulated by principles analogous to those found in the law of nuisance where the "law generally speaks to channeling behavior more than actually prohibiting it. . . . [The] concept of 'indecent' is intimately connected with the exposure of children to language that describes, in terms patently offensive as measured by contemporary community standards for the broadcast medium, sexual or excretory activities and organs, at times of the day when there is a reasonable risk that children may be in the audience."[54]

The Court noted that the FCC, in a follow-up, clarifying opinion said it "never intended to place an absolute prohibition on the broadcast of this type of language, but rather sought to channel it to times of day when children most likely would not be exposed to it."[55] At issue, in part, was the question of what to do about news and other live event coverage where profanity might leak on to the public airwaves.

The FCC had said: " '[In] some cases, public events likely to produce offensive speech are covered live, and there is no opportunity for journalistic editing.' Under these circumstances we believe that it would be inequitable for us to hold a licensee responsible for indecent language. . . . We trust that under such circumstances a licensee will exercise judgment, responsibility, and sensitivity to the community's needs, interests, and tastes."[56] (59 F.C.C. 2d, at 893 n.1.)

The majority took note of the dissent in the opinion by the Court of Appeals: Judge Leventhal, in dissent, stated that the only issue was whether the FCC could regulate the language "as broadcast.". . . Emphasizing the interest in protecting children, not only from exposure to indecent language, but also from exposure to the idea that such language has official approval, . . . he concluded that the FCC had correctly condemned the daytime broadcast as indecent.[57]

The Court took the view that Section 1464 could be squared with the anticensorship language in Section 326 because the FCC's post-broadcast review of programming was not a prior restraint under the First Amendment:

The prohibition against censorship unequivocally denies the Commission any power to edit proposed broadcasts in advance and to excise material considered inappropriate for the airwaves. The prohibition, however, has never been construed to deny the Commission the power to review the content of completed broadcasts in the performance of its regulatory duties.[58]

The legislative history of the Communications Act and later revisions suggested to the Court that Congress had intended that anti-censorship rules and anti-indecency rules could coexist.

Relying upon a dictionary definition (Webster defines the term *indecent* as "a: altogether unbecoming: contrary to what the nature of things or what circumstances would dictate as right or expected or appropriate: hardly suitable: UNSEEMLY . . . b: not conforming to generally accepted standards of morality . . .) the Supreme Court held that the FCC could apply the language in Section 1464:

> The words "obscene, indecent, or profane" are written in the disjunctive, implying that each has a separate meaning. Prurient appeal is an element of the obscene, but the normal definition of "indecent" merely refers to nonconformance with accepted standards of morality.[59]

Would such regulation have a "chilling effect" on broadcasters? The Supreme Court said yes, but argued that the effect on form could be separated from the effect on content:

> It is true that the Commission's order may lead some broadcasters to censor themselves. At most, however, the Commission's definition of indecency will deter only the broadcasting of patently offensive references to excretory and sexual organs and activities. While some of these references may be protected, they surely lie at the periphery of First Amendment concern.[60]

In a footnote of the opinion, the Supreme Court elaborated: "A requirement that indecent language be avoided will have its primary effect on the form, rather than the content, of serious communication. There are few, if any, thoughts that cannot be expressed by the use of less offensive language."[61] The Supreme Court had refused to accept an absolute First Amendment protection for speech—one that relied solely upon the marketplace. In the words of the Court: The words of the Carlin monologue are unquestionably "speech" within the meaning of the First Amendment. It is equally clear that the Commission's objections to the broadcast were based in part on its content. The order must therefore fall if, as

Pacifica argues, the First Amendment prohibits all governmental regulation that depends on the content of speech. Our past cases demonstrate, however, that no such absolute rule is mandated by the Constitution.[62]

The words were offensive, the Court held, in the same way that obscenity offends; to quote Justice Murphy in *Chaplinsky v. New Hampshire*: "[Such] utterances are no essential part of any exposition of ideas, and are of such slight social value as a step to truth that any benefit that may be derived from them is clearly outweighed by the social interest in order and morality."[63]

Likewise, the FCC had reasoned: "Obnoxious, gutter language describing these matters has the effect of debasing and brutalizing human beings by reducing them to their mere bodily functions. . . ."[64] The Supreme Court's *Pacifica* majority added: "Our society has a tradition of performing certain bodily functions in private, and of severely limiting the public exposure or discussion of such matters. Verbal or physical acts exposing those intimacies are offensive irrespective of any message that may accompany the exposure."

Still, even the Carlin words might be seen in context of "social value." In *Cohen v. California*, for example, "Paul Cohen entered a Los Angeles courthouse wearing a jacket emblazoned with the words "Fuck the Draft." After entering the courtroom, he took the jacket off and folded it; . . . evidence showed, no one in the courthouse was offended by his jacket. Nonetheless, when he left the courtroom, Cohen was arrested, convicted of disturbing the peace, and sentenced to 30 days in prison."[65]

> In holding that criminal sanctions could not be imposed on Cohen for his political statement in a public place, the Court rejected the argument that his speech would offend unwilling viewers; it noted that "there was no evidence that persons powerless to avoid [his] conduct did in fact object to it." In contrast, in this case the Commission was responding to a listener's strenuous complaint, and Pacifica does not question its determination that this afternoon broadcast was likely to offend listeners. It should be noted that the Commission imposed a far more moderate penalty on Pacifica than the state court imposed on Cohen. Even the strongest civil penalty at the Commission's command does not include criminal prosecution.[66]

In *Pacifica*, the Supreme Court recognized that broadcasting is a unique medium in terms of access afforded children, and in terms of their ability to use electronic media: "Although Cohen's written message might have been incomprehensible to a first grader, Pacifica's broadcast could have enlarged a child's vocabulary in an instant."[67] And even these restrictions, the Court held, could be seen as appropriate in light of the restrictions we place on access to other forms of media. The Supreme Court noted:

> The Commission's action does not by any means reduce adults to hearing only what is fit for children. . . . Adults who feel the need may purchase tapes and records or go to theaters and nightclubs to hear these words. In fact, the Commission has not unequivocally closed even broadcasting to speech of this sort; whether broadcast audiences in the late evening contain so few children that playing this monologue would be permissible is an issue neither the Commission nor this Court has decided.[68]

It is important to note that the Court, in emphasizing "the narrowness of our holding," appeared to conclude that the regulation of indecent speech over the public airwaves would not do damage to serious First Amendment concerns:

> The concept requires consideration of a host of variables. The time of day was emphasized by the Commission. The content of the program in which the language is used will also affect the composition of the audience,** and differences between radio, television, and perhaps closed-circuit transmissions, may also be relevant. As Mr. Justice Sutherland wrote, a "nuisance may be merely a right thing in the wrong place—like a pig in the parlor instead of the barnyard." . . . We simply hold that when the Commission finds that a pig has entered the parlor, the exercise of its regulatory power does not depend on proof that the pig is obscene.[69]** Even a prime-time recitation of Geoffrey Chaucer's *Miller's Tale* would not be likely to command the attention of many children who are both old enough to understand and young enough to be adversely affected by passages such as: "And prively he caughte hire by the queynte."[70]

The three-member majority opinion was joined in a concurrence by Justice Powell, joined by Justice Blackmun to form the 5-4 decision of the Court.

We now turn to the eloquent dissent of Justice Brennan that was joined by Justice Marshall, who wrote: "The language involved in this case is as potentially degrading and harmful to children as representations of many erotic acts.[71] The opinion is reproduced, in part, below:

Federal Communications Commission v. Pacifica Foundation

438 U.S. 726 (1978)

Dissent by Justice Brennan,

[T]he word "indecent" . . . must be construed to prohibit only obscene speech. I would, therefore, normally refrain from expressing my views on any constitutional issues implicated in this case. However, I find the Court's misapplication of fundamental First Amendment principles so patent, and its attempt to impose its notions of propriety on the whole of the American people so misguided, that I am unable to remain silent.

[A]ll Members of the Court agree that the Carlin monologue aired by Station WBAI does not fall within one of the categories of speech, such as "fighting words," . . . or obscenity, . . . that is totally without First Amendment protection. This conclusion, of course, is compelled by our cases expressly holding that communications containing some of the words found condemnable here are fully protected by the First Amendment in other contexts. . . .

This majority apparently believes that the FCC's disapproval of Pacifica's afternoon broadcast of Carlin's "Dirty Words" recording is a permissible time, place, and manner regulation. . . . Both the opinion of my Brother STEVENS and the opinion of my Brother POWELL rely principally on two factors in reaching this conclusion: (1) the capacity of a radio broadcast to intrude into the unwilling listener's home, and (2) the presence of children in the listening audience. Dispassionate analysis, removed from individual notions as to what is proper and what is not, starkly reveals that these justifications, whether individually or together, simply do not support even the professedly moderate degree of governmental homogenization of radio communications—if, indeed, such homogenization can ever be moderate given the pre-eminent status of the right of free speech in our constitutional scheme—that the Court today permits.

Rejecting an argument that privacy rights justify banning indecency, Justice Brennan said the Court misunderstood the valued right of privacy.

To reach a contrary balance, as does the Court, is clearly to follow MR. JUSTICE STEVENS' reliance on animal metaphors, . . . "to burn the house to roast the pig." The Court's balance, of necessity, fails to accord proper weight to the interests of listeners who wish to hear broadcasts the FCC deems offensive. It permits majoritarian tastes completely to preclude a protected message from entering the homes of a receptive, unoffended minority. No decision of this Court supports such a result. Where the individuals constituting the offended majority may freely choose to reject the material being offered, we have never found their privacy interests of such moment to warrant the suppression of speech on privacy grounds. . . .

Most parents will undoubtedly find understandable as well as commendable the Court's sympathy with the FCC's desire to prevent offensive broadcasts from reaching the ears of unsupervised children. Unfortunately, the facial appeal of this justification for radio censorship masks its constitutional insufficiency. Although the government unquestionably has a special interest in the well-being of children and consequently "can adopt more stringent controls on communicative materials available to youths than on those available to adults" . . .

Because the Carlin monologue is obviously not an erotic appeal to the prurient interests of children, the Court, for the first time, allows the government to prevent minors from gaining access to materials that are not obscene, and are therefore protected, as to them. It thus ignores our recent admonition that "[speech] that is neither obscene as to youths nor subject to some other legitimate proscription cannot be suppressed solely to protect the young from ideas or images that a legislative body thinks unsuitable for them."

In concluding that the presence of children in the listening audience provides an adequate basis for the FCC to impose sanctions for Pacifica's broadcast of the Carlin monologue, the opinions of my

Brother POWELL, . . . and my Brother STEVENS, . . . both stress the time-honored right of a parent to raise his child as he sees fit—a right this Court has consistently been vigilant to protect.

As surprising as it may be to individual Members of this Court, some parents may actually find Mr. Carlin's unabashed attitude towards the seven "dirty words" healthy, and deem it desirable to expose their children to the manner in which Mr. Carlin defuses the taboo surrounding the words. Such parents may constitute a minority of the American public, but the absence of great numbers willing to exercise the right to raise their children in this fashion does not alter the right's nature or its existence. Only the Court's regrettable decision does that.

Taken to their logical extreme, these rationales would support the cleansing of public radio of any "four-letter words" whatsoever, regardless of their context. The rationales could justify the banning from radio of a myriad of literary works, novels, poems, and plays by the likes of Shakespeare, Joyce, Hemingway, Ben Jonson, Henry Fielding, Robert Burns, and Chaucer; they could support the suppression of a good deal of political speech, such as the Nixon tapes; and they could even provide the basis for imposing sanctions for the broadcast of certain portions of the Bible.

(Note 5: See, e.g., I Samuel 25:22: "So and more also do God unto the enemies of David, if I leave of all that pertain to him by the morning light any that pisseth against the wall"; II Kings 18:27 and Isaiah 36:12: "[Hath] he not sent me to the men which sit on the wall, that they may eat their own dung, and drink their own piss with you?"; Ezekiel 23:3: "And they committed whoredoms in Egypt; they committed whoredoms in their youth; there were their breasts pressed, and there they bruised the teats of their virginity."; Ezekiel 23:21: "Thus thou calledst to remembrance the lewdnes of thy youth, in bruising thy teats by the Egyptians for the paps of thy youth." The Holy Bible [King James Version][Oxford 1897].)

In order to dispel the specter of the possibility of so unpalatable a degree of censorship, and to defuse Pacifica's overbreadth challenge, the FCC insists that it desires only the authority to reprimand a broadcaster on facts analogous to those present in this case, which it describes as involving "broadcasting for nearly twelve minutes a record which repeated over and over words which depict sexual or excretory activities and organs in a manner patently offensive by its community's contemporary standards in the early afternoon when children were in the audience" . . . The opinions of both my Brother POWELL and my Brother STEVENS take the FCC at its word, and consequently do no more than permit the Commission to censor the afternoon broadcast of the "sort of verbal shock treatment" . . .

It is quite evident that I find the Court's attempt to unstitch the warp and woof of First Amendment law in an effort to reshape its fabric to cover the patently wrong result the Court reaches in this case dangerous as well as lamentable. Yet there runs throughout the opinions of my Brothers POWELL and STEVENS another vein I find equally disturbing: a depressing inability to appreciate that in our land of cultural pluralism, there are many who think, act, and talk differently from the Members of this Court, and who do not share their fragile sensibilities. It is only an acute ethnocentric myopia that enables the Court to approve the censorship of communications solely because of the words they contain. "A word is not a crystal, transparent and unchanged, it is the skin of a living thought and may vary greatly in color and content according to the circumstances and the time in which it is used" . . . The words that the Court and the Commission find so unpalatable may be the stuff of everyday conversations in some, if not many, of the innumerable subcultures that compose this Nation. Academic research indicates that this is indeed the case. As one researcher concluded, "[words] generally considered obscene like 'bullshit' and 'fuck' are considered neither obscene nor derogatory in the [black] vernacular except in particular contextual situations and when used with certain intonations."

Today's decision will thus have its greatest impact on broadcasters desiring to reach, and listening audiences composed of, persons who do not share the Court's view as to which words or expressions are acceptable and who, for a variety of reasons, including a conscious desire to flout majoritarian conventions, express themselves using words that may be regarded as offensive by those from different socioeconomic backgrounds. In this context, the Court's decision may be seen for what, in the broader perspective, it really is: another of the dominant culture's inevitable efforts to force those groups who do not share its mores to conform to its way of thinking, acting, and speaking.

Justice Stewart wrote a separate dissent in *Pacifica* that was joined by Justices Brennan, White, and Marshall. Significant in that view was that broadcasters would be, under the FCC's regulation, held to a stricter standard than found in obscenity law: I think that "indecent" should properly be read as meaning no more than "obscene." Since the Carlin monologue concededly was not "obscene," I believe that the FCC lacked statutory authority to ban it.[72]

The facts, legal issues, judicial reasoning, and holdings of *Pacifica*

The briefing method used commonly by law students is a clean way to summarize a complicated opinion.

Facts: A New York radio station had broadcast the Carlin monologue "Filthy Words" at 2 P.M., and there was a single complaint from a man who said his "young son" heard the profanity. Following FCC inquiry, the station defended the broadcast. The FCC placed a note in the station's file, and that action was challenged in court. The United States Court of Appeals, District of Columbia Circuit, ruled in favor of Pacifica, but the Supreme Court reversed.

Issues: Is 18 U.S.C. 1464 consistent with the First Amendment and Section 326 of the Communications Act of 1934? Is broadcasting uniquely accessible by young children requiring special regulation? Are there privacy issues involved in the broadcast of radio into the home? Does the public interest standard allow for media content based regulation?

Reasoning: The majority opinion held: "A requirement that indecent language be avoided will have its primary effect on form, rather than content, of serious communication. There are few, if any, thoughts that cannot be expressed by the use of less offensive language."

A dissenting view was, "As surprising as it may be to individual Members of this Court, some parents may actually find Mr. Carlin's unabashed attitude towards the seven 'dirty words' healthy, and deem it desirable to expose their children to the manner in which Mr. Carlin defuses the taboo surrounding the words. Such parents may constitute a minority of the American public, but

---

the absence of great numbers willing to exercise the right to raise their children in this fashion does not alter the right's nature or its existence."

Decision: The Court, on a 5-4 vote reversed the appellate court and reinstated the FCC's right to regulate indecent speech on radio.

Manager's Summary

Broadcast managers looking for a "bottom line" to this chapter should consider two main points:

1. The law of indecency is that broadcasters do face legal limits as to "indecent" content programmers might want to use. First Amendment protections are not unlimited.
2. As a practical matter, common sense will keep a broadcaster out of trouble with the FCC. An honest, good faith approach to following the regulations can allow broadcasters to assert strong free speech rights.

Notes

1. See Paul D. Driscoll, "The Federal Communications Commission and Broadcast Indecency," Law Division, Association for Education in Journalism and Mass Communication, April 1989. Cf. Theodore L. Glasser, "The Press, Privacy, and Community Mores," Mass Communication Division, Speech Communication Association, Louisville, Kentucky, November 1982; and Matthew L. Spitzer, "Controlling the Content of Print and Broadcasting," 58 S. Cal. L. Rev. 1349 (1985), *Seven Dirty Words and Six Other Stories: Controlling the Content of Print and Broadcast*, New Haven: Yale University Press, 1986. Cf. Timothy B. Dyk, book review, 40(1) *Federal Comm. L. J.* 131–41 (1988).
2. *Yale Broadcasting v. FCC*, 478 F.2d 594 (D.C. Cir. 1973).
3. *Pacifica*, at fn. 13.
4. *Pacifica Foundation*, 36 FCC 147 (1964).
5. *In Re WUHY-FM Eastern Educational Radio*, 24 FCC 2d 408 (1970).
6. *In Re Apparent Liability, WGLD-FM*, 41 F.C.C. 2d 919 (1973); Sonderling, 41 F.C.C. 2d 777 (1973); *Illinois Citizens Committee for Broadcasting v. FCC*, 515 F. 2d 397 (D.C. Cir. 1974).

7. See Krasnow, Longley, and Terry (1982) for a discussion of how failed policy may result from enforcement based upon political factors. For more on the general regulatory problem, see Michael Starr and David Atkin, "The Department of Communications: A Plan for the Abolition of the Federal Communications Commission," Radio-Television Journalism Division, Association for Education in Journalism and Mass Communication, national conference, Washington, D.C., August 1989. For a sense of the political climate on indecency, one only need read the comments of Senator Strom Thurmond: "It is readily apparent that through thinly veiled innuendo or with unbridled candor, sexually explicit material is growing by leaps and bounds. . . ." 135 Cong. Rec. S. 12928 (18 September 1989).

8. *Infinity Broadcasting*, 2 FCC Rcd. 2705 (1987). Howard Stern, the most well-known of shock jocks, was cited for talking about "testicles," "homos," "lesbians," and "sodomy." Stern's dialogue avoids *Pacifica* Carlin monologue words.

9. Senator Albert Gore (D-Tennessee) challenged the nomination of Commissioner Andrew C. Barrett, who had said, "[T]he FCC must recognize 'there is a market for indecency' and the agency's role is to decide 'what the law will allow us to tolerate.' " Gore responded that Americans are "sick and tired of what's going on." Quoted at, 135 Cong Rec S 10384, *S10386.

10. Howard M. Kleiman, "Indecent Programming on Cable Television: Legal and Social Dimensions," *Journal of Broadcasting & Electronic Media* 30(3):275–94 (Summer 1986).

11. In *Red Lion Broadcasting v. FCC*, 395 U.S. 367 (1969), there is an argument that listeners' and viewers' rights are "paramount" to the First Amendment rights of broadcasters; and *NBC v. United States*, 319 U.S. 190 (1943), a classic defense of the scarcity rationale, positions the FCC as more than simply a "traffic cop"—the agency is charged with the responsibility to monitor traffic composition, namely content.

12. Jeremy Harris Lipschultz, " 'Political Propaganda': The Supreme Court Decision in *Meese v. Keene*," *Communications and the Law* 11(4):25–44 (December 1989).

13. 438 U.S. 729–31.

14. See, *Telecommunications Research and Action Center v. FCC*, 800 F.2d 1181 (D.C. Cir. 1986), for the view that technical or physical scarcity no longer justifies regulation of the broadcast medium, or treating it differently from the printed press in terms of the first amendment.

15. 438 U.S. 743, fn. 18 (Opinion of Stevens, J.). A twenty-four-hour ban, as an absolute prohibition, could be reasoned to be "censorship" not narrowly crafted. See the test of *United States v. O'Brien*, 391 U.S. 367 (1968). A governmental restriction should be "no greater than essential" to further a substantial interest.

16. 438 U.S. 770, 773 (Brennan, J. dissenting).

17. *New Indecency Enforcement Standards*, 2 FCC Rcd. 2726 (1987). The FCC broadened its approach, moving away from the Carlin words to a "generic" approach in the definition of broadcast indecency: "Language or material that depicts or describes, in terms patently offensive as measured by contemporary community standards for the broadcast medium, sexual or excretory activities

or organs." The FCC also found reasonable risk that children would be in the audience after 10 P.M., the previously protected time. Cf. *Pacifica Foundation*, 2 FCC Rcd. 2698 (1987); and *Regents of the University of California*, 2 FCC Rcd. 2703 (1987).

18. *In the Matter of Enforcement of Prohibitions Against Broadcast Obscenity and Indecency in 18 U.S.C. Sec. 1464*, Order, FCC 88-416 (Dec. 19, 1988). Also see 53(249) FR 52425 (Dec. 28, 1988).
19. FCC 88–416, point number 3.
20. Separate Statement of Commissioner Patricia Diaz Dennis (Dec. 12, 1989).
21. *Sable Communications v. FCC*, 109 S.Ct. 2829 (1989).
22. 413 U.S. 49, 103 (1973).
23. See *Broadcasting*, "FCC Cleans Out the Pipeline on Indecency," 30 October 1989, p. 28; and *Broadcasting*, "Radio Broadcasters Troubled by Sikes FCC 'Moving the Goal Posts' on Indecency," 6 November 1989, pp. 66–67. One broadcaster said, "I don't know how any group in Washington can set themselves up as defining community standards in each market in the nation."
24. The information was obtained through a *Freedom of Information Act* (5 U.S.C. Section 552) request under FOIA Control No. 89-217, in a packet dated February 8, 1990. A request to obtain copies of station response letters is pending in a request letter dated 9 March 1990. For background on the problems with FOIA requests to the FCC, see *Media Access Project*, 41 FCC 2d 179 (1973), and *In the Matter of Rob Warden*, 70 FCC 2d 1735 (1978).
25. FCC letter, 8310-TRW, 24 August 1989.
26. 18 U.S.C. 1464 reads: "Whoever utters any obscene, indecent or profane language by means of radio communication shall be fined not more than $10,000 or imprisoned not more than two years, or both." The FCC cites the language of 47 U.S.C. Sections 312(a)(6) and 503(b)(1)(D) as its statutory authority. to punish licensees. The Section 312 provisions allow the FCC to "revoke any station license or construction permit" for lottery (Section 1304), fraud (Section 1343), or indecency (Section 1464) reasons. Section 503 provides for "forfeitures"—fines—for violations. Enforcement of Section 1464 in terms of a criminal penalty would be a matter for the Justice Department, but no known cases exist.
27. The FCC uses as its authority: *FCC v. Pacifica Foundation*, 438 U.S. 726 (1978); *Pacifica Foundation, Inc.*, 2 FCC Rcd 2698 (1987); *The Regents of the University of California*, 2 FCC Rcd 2703 (1987), and *Infinity Broadcasting Corp. of Pa.*, 2 FCC Rcd 2705 (1987), *Order on reconsideration, Infinity Broadcasting Corp.*, 3 FCC Rcd 930 (1987) *("Reconsideration Order"), aff'd in part and remanded in part sub. nom. Action for Children's Television v. FCC*, 852 F.2d 1332 (D.C. Cir. 1988) *("ACT I")*. The *ACT I* decision was treated as "governing law" by the FCC while review on the twenty-four-hour ban continued.
28. 8310-MD, November 30, 1989 letter to Evergreen Media.
29. Ibid., at 1.
30. Ibid., at 2.
31. *Broadcasting*, editorial, "Bad to Worse," 30 October 1989.
32. National Association of Broadcasters, *TV Today*, 26 February 1990.
33. Radio-Television News Directors Association, "RTNDA Joins Appeal for Review of Indecency Rule," *Intercom* 6(2):2 (19 January 1989).

34. *Broadcasting*, "Chances Slim for 24-Hour FCC Ban on Indecency," 30 January 1989.
35. *Broadcasting*, Law and Regulation column, "FCC Hears Little Support for 24-Hour Broadcasting Indecency Ban," 26 February 1990.
36. 24 FCC 2d 408 (1970).
37. 41 FCC 2d 919 (1973).
38. Ibid.
39. 41 FCC 2d 777 (1973).
40. 515 F.2d 397 (1975).
41. 478 F.2d 594 (1973).
42. 51 FCC 2d 418 (1975).
43. 57 FCC 2d 782 (1975).
44. 2 FCC Rcd. 2705 (1987).
45. 852 F.2d 1332, 1340 (D.C. Cir. 1988): "Indecent but not obscene material, we reiterate, qualifies for first amendment protection whether or not it has serious merit."
46. Ibid., at 1344, fn. 21: "Petitioners argue cogently parental authority is enhanced, not reduced if government permits programming at hours outside the workday hours common in the community, when most parents can supervise their children's listening.
47. In *Infinity* the Commission found "untenable the view that the holding in *Pacifica* limits a finding of indecency to the use of the seven offensive words contained in the Carlin monologue." It is notable that popular media complaints about Howard Stern date to at least 1984, as evidenced by a *People* magazine piece of that year. It is not clear that *Pacifica* caused a lull in indecency actions against stations; it may have been FCC reluctance to enforce the rules until the FCC began to feel political pressure from the Congress.
48. See, for example, Paul Messaris and Dennis Kerr, "TV-Related Mother-Child Interaction and Children's Perceptions of TV Characters," *Journalism Quarterly* 61(3):662–66 (Autumn 1984): In the case of televised aggression and family environment, "information on the consequences of such behavior is still lacking." It is unclear how indecent language, presumably more subtle than violence or obscenity, can have a negative effect on cognitive or affective development.
49. Richard G. Passler, "Comment: Regulation of Indecent Radio Broadcasts: George Carlin Revisited—What Does the Future Hold for the Seven 'Dirty' Words?" 65 *Tulane Law Review* 131 (November 1990), citing "Filthy Words, Occupation: Foole," (Little David Records 1973).
50. Ibid., note 12, citing *Broadcasting*, 10 July 1978, at 20.
51. Carter, Franklin, and Wright, p. 67.
52. Ibid., p. 66, citing *National Broadcasting Co. v. United States*, 319 U.S. 190 (1943).
53. Ibid., p. 77, citing *Red Lion Broadcasting Co. v. Federal Communications Commission*, 395 U.S. 367 (1969).
54. *Pacifica*, 438 U.S., at pp. 731–732.
55. Ibid., citing 59 F.C.C. 2d 892 (1976).
56. Ibid., 438 U.S., at p. 733.
57. Ibid., at p. 734.
58. Ibid., at p. 735.

59. Ibid., at pp. 739–740.
60. Ibid., at p. 743.
61. Ibid., note 18.
62. Ibid., at p. 744.
63. Ibid., at p. 746.
64. Ibid., note 23.
65. Ibid., p. 747, at note 25.
66. Ibid., second paragraph of note 25.
67. Ibid., p. 749.
68. Ibid., p. 750, note 28.
69. Ibid., p. 750, above.
70. Ibid., p. 750, note 29.
71. Ibid., pp. 755–762, 757: "But it is also true that the language employed is, to most people, vulgar and offensive. It was chosen specifically for this quality, and it was repeated over and over as a sort of verbal shock treatment. The Commission did not err in characterizing the narrow category of language used here as 'patently offensive' to most people regardless of age."
72. Ibid., p. 778.

Chapter 3
Origins of the Concept of "Indecent" Communication

"If prostitution is the world's oldest profession," writes William Layman, "then perhaps pornography is the world's oldest expression."[1] According to Layman:

> The value of expressing the pornographic, obscene, violent, sexual, or repulsive has long been recognized by comedians, lovers, artists, advertisers, and entrepreneurs. Setting aside the definitional problem of what obscenity and pornography actually are, it is undeniable that the qualities which suggest the obscene or pornographic—some degree of sexual suggestiveness, violence, interest in excretory functions, animalism, domination, subordination, or inequality—form a significant portion of our daily diet of expression, and always have.[2]

By one account, "Pornographic sketches done by primitive Homosapiens have been found on the walls of ancient caves," and pornographic etchings have also been uncovered on the walls of Pompeii dating back two thousand years.[3] Layman continues: "For artists the value of expressing the obscene or sexual seems to arise from the very fact that these forms of expression have been repressed and mystified over the ages."[4]

The Common Law of Obscenity: Historical Notes

In the view of Gillmor, Baron, Simon, and Terry (1990), the English courts developed the common law of obscenity in response to the mores of the eighteenth century:

The time was ripe. Obscenity, and vice societies bent on stamping it out, were both gaining momentum. By the beginning of the nineteenth century, England had entered a period of sexual explicitness.[5]

The landmark *R. v. Hicklin* (1868) case seemed to target protection of "the most feeble-minded and susceptible persons." An obscene book was seen by Lord Chief Justice Cockburn in the Court of Queen's Bench as a danger to some readers: "Whether the tendency of the matter charged as obscenity is to deprave and corrupt those whose minds are open to such immoral influence and into whose hands a publication of this sort may fall." The first American attack on obscenity appears to have come as the Tarriff Act of 1842 sought to restrict importation of those European materials deemed as obscene.[6]

Mass media historian and legal scholar Margaret Blanchard writes that Americans have had and continue to exhibit a history of tolerance for suppression of objectionable speech:

> [M]any of them have become increasingly willing to allow the government to intrude into their leisure time activities in an effort to cleanse society from excessive sexuality and to protect children from the perverting influences of various media forms. No longer is the family considered able or, perhaps more accurately, willing to set standards of behavior for its members. Rather than simply forbidding young people to listen to certain forms of music, read certain books, or see certain movies, many families have abdicated this responsibility to civic action groups and the government. Such a relinquishment of authority over individual lives has led to denunciations of various media forms, calls for self-regulation of individual mediums, and attempts to ban completely some sexually explicit speech.[7]

Her study of the campaigns dating to the last quarter of the nineteenth century reveal the following generalizations:

(1)   Conservative trends in political and economic life are strongly connected with such clean-up campaigns. Indeed, political conservatives may encourage attacks on sexually explicit materials in an effort to divert American energy from areas in which it could cause trouble for conservative interests.

(2)   Large-scale attacks generally begin with criticism of fringe materials in which few can find redeeming social value. Ultimately,

however, the campaigns to clean up society try to expunge materials that most Americans consider important or valuable. For example, before he finished his career, religious crusader Anthony Comstock attacked nude paintings by modern French masters and a play by George Bernard Shaw. Early crusaders also helped retard the distribution of information on sex education, birth control, and abortion.

(3) The media almost invariably yield to pressure from the attack and establish some sort of code of self-regulation to keep the reformers at bay. This was as true of the dime novels in Anthony Comstock's time as of the record industry today.

(4) Parents and grandparents who lead the efforts to cleanse today's society seem to forget that they survived alleged attacks on their morals by different media when they were children. Each generation of adults either loses faith in the ability of its young people to do the same or becomes convinced that the dangers facing the new generation are much more substantial than the ones it faced as children.

(5) The support for sexually explicit expression has never been strong. Most people simply do not want to talk about such materials. The words of George Carlin's infamous monologue or 2 Live Crew's obnoxious lyrics are conspicuously absent from major media sources. In almost every instance in which sexually explicit material is threatened, some such support does appear, however reluctantly, but convincing people that this form of speech deserves protection is most difficult.

(6) In asking federal, state, and local governments to take action against sexually explicit speech, Americans are requesting intervention in the most private areas of family life—the right to inculcate in their children the moral values that they wish to pass on. Legislators, activists, and judges are making more of these decisions than ever before, and their standards may well not be those desired by individual families.[8]

Blanchard found that one early crusade had been successful in attacking both low and high literature. The Comstock Act (driven by the crusading Anthony Comstock) specified that "no obscene, lewd, or lascivious book, pamphlet, picture, paper, print, or other publication of an indecent character, or any article or thing designed or intended for the prevention of conception or procuring of abortion, nor any article or thing intended or adapted for any indecent or immoral use or nature . . . shall be carried in the mail."[9] Blanchard found that the *Hicklin* test was used by the courts to interpret the Comstock Act strictly:

> Judges using this standard condemned books because parts of them might be considered obscene by the young and inexperienced people who might happen to read them rather than by evaluating the books by the standards of the intended audience. Court officials refused to enter obscene material into the record for fear of offending persons attending the session, and this further enhanced Comstock's ability to obtain convictions. In addition, jurors were not allowed to hear so-called expert witnesses testify to the value of the material being challenged. The court system was therefore fairly well-rigged to guarantee that material challenged as obscene would be found to be so.[10]

We can see in Comstock's language from an earlier time a common thread of concern that persists today in the broadcast indecency arena: "[A] man may think, write, and speak as he pleases by himself, but he must put his public utterances into decent language."[11] Anthony Comstock's campaign eventually extended to the dime novel, which was connected with antisocial and criminal behavior, and the publishers took steps to self-regulate to protect their businesses.[12]

Such morality campaigns can be seen as seminal in the later efforts to control what children were exposed to in movies, comic books, and rock music.[13] As such, the cases throughout this century have used the term "indecent" loosely, and courts have been willing to associate such matters with obscenity—an area lacking constitutional protection.

The significance for broadcasters is that questions of morality have been on the regulatory table for years, but the FCC has mostly left it to stations to exercise self-regulation. As mass media Professor John Bittner notes, the FCC has exercised little control over the content of messages broadcast: "With the exception of obscene and indecent programming—and even that area is somewhat nebulous—lotteries and advertising are about the only areas of programming the FCC can directly regulate without infringing on the First Amendment."[14]

Miller **and** *Pope* **Define Obscenity**

The definition of "indecent" seems to range from simple profanity, as outlined in the *Pacifica* case, to later attempts by the FCC to judge sexual innuendo, to being synonymous with pornographic obscenity. As such, we need to examine what the United States Supreme Court has said about obscenity in the print media context. We'll look at portions of three of the opinions in *Pope v. Illinois*, a 1987 case that shows the Court itself has become troubled by the inability to define legal concepts:

POPE ET AL. v. ILLINOIS

481 U.S. 497, 107 S. Ct. 1918,

95 L. Ed. 2d 439, 4 Media L. Rep. 1001 (1987)

(Justice White delivered the opinion of the Court, in which Justices Rehnquist, Powell, O'Connor, and Scalia, joined, in part. Justice Scalia filed a concurring opinion, and Justice Blackmun concurred in part and dissented in part. Justice Brennan dissented. Justice Stevens filed a dissenting opinion, in which Justices Marshall, Brennan, and Blackman joined in part.)

Justice White's Majority Opinion said:

In *Miller v. California* (1973), the Court set out a tripartite test for judging whether material is obscene. The third prong of the *Miller* test requires the trier of fact to determine "whether the work, taken as a whole, lacks serious literary, artistic, political, or scientific value." The issue in this case is whether, in a prosecution for the sale of allegedly obscene materials, the jury may be instructed to apply community standards in deciding the value question.

On July 21, 1983, Rockford, Illinois, police detectives purchased certain magazines from the two petitioners, each of whom was an attendant at an adult bookstore. Petitioners were subsequently charged separately with the offense of "obscenity" for the sale of these magazines. Each petitioner moved to dismiss the charges against him on the ground that the then-current version of the Illinois obscenity statute, ... (1983), violated the First and Fourteenth Amendments to the United States Constitution. Both

petitioners argued, among other things, that the statute was unconstitutional in failing to require that the value question be judged
"solely on an objective basis as opposed to reference [sic] to contemporary community standards." Both trial courts rejected this contention and instructed the respective juries to judge whether the
material was obscene by determining how it would be viewed by
ordinary adults in the whole State of Illinois.

There is no suggestion in our cases that the question of the value of an
allegedly obscene work is to be determined by reference to community standards. Indeed, our cases are to the contrary. *Smith v. United
States*, . . . (1977), held that, in a federal prosecution for mailing
obscene materials, the first and second prongs of the *Miller* test—
appeal to prurient interest and patent offensiveness—are issues of
fact for the jury to determine applying contemporary community
standards. The Court then observed that, unlike prurient appeal and
patent offensiveness, "literary, artistic, political, or scientific value . . .
is not discussed in *Miller* in terms of contemporary community standards." [T]he Court was careful to point out that "the First
Amendment protects works which, taken as a whole, have serious literary, artistic, political, or scientific value, regardless of whether the
government or a majority of the people approve of the ideas these
works represent" . . . Just as the ideas a work represents need not
obtain majority approval to merit protection, neither, insofar as the
First Amendment is concerned, does the value of the work vary from
community to community based on the degree of local acceptance it
has won. The proper inquiry is not whether an ordinary member of
any given community would find serious literary, artistic, political, or
scientific value in allegedly obscene material, but whether a reasonable person would find such value in the material, taken as a whole.
The instruction at issue in this case was therefore unconstitutional.

Justice Scalia, Concurring.

I join the Court's opinion with regard to harmless error because I
think it implausible that a community standard embracing the
entire State of Illinois would cause any jury to convict where a "reasonable person" standard would not. At least in these circumstances, if a reviewing court concludes that no rational juror,
properly instructed, could find value in the magazines, the
Constitution is not offended by letting the convictions stand. I join

the Court's opinion with regard to an "objective" or "reasonable person" test of "serious literary, artistic, political, or scientific value," *Miller v. California* (1973), because I think that the most faithful assessment of what *Miller* intended, and because we have not been asked to reconsider *Miller* in the present case. I must note, however, that in my view it is quite impossible to come to an objective assessment of (at least) literary or artistic value, there being many accomplished people who have found literature in Dada, and art in the replication of a soup can. Since ratiocination has little to do with esthetics, the fabled "reasonable man" is of little help in the inquiry, and would have to be replaced with, perhaps, the "man of tolerably good taste"—a description that betrays the lack of an ascertainable standard. If evenhanded and accurate decision making is not always impossible under such a regime, it is at least impossible in the cases that matter. I think we would be better advised to adopt as a legal maxim what has long been the wisdom of mankind: *De gustibus non est disputandum.* Just as there is no use arguing about taste, there is no use litigating about it.

For the law courts to decide "What is Beauty" is a novelty even by today's standards. . . . It is a refined enough judgment to estimate whether a reasonable person would find literary or artistic value in a particular publication; it carries refinement to the point of meaninglessness to ask whether he could do so. Taste being, as I have said, unpredictable, the answer to the question must always be "yes" . . .

Justice Brennan, Dissenting.

Justice Stevens persuasively demonstrates the unconstitutionality of criminalizing the possession or sale of "obscene" materials to consenting adults. I write separately only to reiterate my view that any regulation of such material with respect to consenting adults suffers from the defect that "the concept of 'obscenity' cannot be defined with sufficient specificity and clarity to provide fair notice to persons who create and distribute sexually oriented materials, to prevent substantial erosion of protected speech as a byproduct of the attempt to suppress unprotected speech, and to avoid very costly institutional harms" . . .

Justice Stevens, . . .

The Court correctly holds that the juries that convicted petitioners were given erroneous instructions on one of the three essential elements of an obscenity conviction. Nevertheless, I disagree with its disposition of the case for three separate reasons: (1) the error in the instructions was not harmless; (2) the Courts attempt to clarify the constitutional definition of obscenity is not faithful to the First Amendment; and (3) I do not believe Illinois may criminalize the sale of magazines to consenting adults who enjoy the constitutional right to read and possess them.

The distribution of magazines is presumptively protected by the First Amendment. The Court has held, however, that the constitutional protection does not apply to obscene literature. If a state prosecutor can convince the trier of fact that the three components of the obscenity standard set forth in *Miller v. California* (1973) are satisfied, it may, in the Court's view, prohibit the sale of sexually explicit magazines. In a criminal prosecution, the prosecutor must prove each of these three elements beyond a reasonable doubt. Thus, in these cases, in addition to the first two elements of the *Miller* standard, the juries were required to find, on the basis of proof beyond a reasonable doubt, that each of the magazines "lacks serious literary, artistic, political, or scientific value."

The required finding is fundamentally different from a conclusion that a majority of the populace considers the magazines offensive or worthless. As the Court correctly holds, the juries in these cases were not instructed to make the required finding; instead, they were asked to decide whether "ordinary adults in the whole State of Illinois" would view the magazines that petitioners sold as having value.

There is an even more basic reason why I believe these convictions must be reversed. The difficulties inherent in the Court's "reasonable person" standard reaffirm my conviction that government may not constitutionally criminalize mere possession or sale of obscene literature, absent some connection to minors or obtrusive display to unconsenting adults. During the recent years in which the Court has struggled with the proper definition of obscenity, six Members of the Court have expressed the opinion that the First Amendment, at the very least, precludes criminal prosecutions for sales such as those involved in this case. Dissenting in *Smith v. United States* (1977), I explained my view: "The question of offensiveness to community standards, whether national or local, is not one that the average juror can be expected to answer with evenhanded consistency. The average juror may well have one reaction to sexually oriented materials in a completely private setting and an entirely different reaction in a social context. Studies have shown that an opinion held by a large majority of a group concerning a neutral and objective subject has a significant impact in distorting the perceptions of group members who would normally take a different position.

Since obscenity is by no means a neutral subject, and since the ascertainment of a community standard is such a subjective task, the expression of individual jurors' sentiments will inevitably influence the perceptions of other jurors, particularly those who would normally be in the minority. Moreover, because the record never discloses the obscenity standards which the jurors actually apply, their decisions in these cases are effectively unreviewable by an appellate court. In the final analysis, the guilt or innocence of a criminal defendant in an obscenity trial is determined primarily by individual jurors' subjective reactions to the materials in question rather than by the predictable application of rules of law. "This conclusion is especially troubling because the same image—whether created by words, sounds, or pictures—may produce such a wide variety of reactions. As Mr. Justice Harlan noted: '[It is] often true that one man's vulgarity is another's lyric. Indeed, we think it is largely because government officials [or jurors] cannot make principled distinctions in this area that the Constitution leaves matters of taste and style so largely to the individual' . . . In my judgment, the line between communications which 'offend' and those which do not is too blurred to identify criminal conduct. It is also too blurred to delimit the protections of the First Amendment."

I respectfully dissent.

The *Pope* decision makes clear that the regulation of obscenity is, most often, grounded in the desire to protect: (1) children; and (2) unconsenting adults. This justification is identical to that found in the broadcasting *Pacifica* case, where the issue is punishment for indecency over the broadcast airwaves.

A key question is whether the lack of an ability to define obscenity, as Justice Scalia notes, has any bearing on the similar problems associated with definitions of broadcast indecency. We will explore this question as we see how the FCC and courts have interpreted the definitional crisis in the law.

Indecency Applications

Middleton and Chamberlin are among those taking a strong free speech position on indecency: "Indecency is protected by the First Amendment because the Supreme Court has said that only those materials meeting the *Miller v. California* tests fall outside of First Amendment protection."[15] Still, it is the *Pacifica* ruling that introduced the "however" to Miller.[16] As we will see later in this book, lower courts have upheld the FCC's position that broadcast indecency is outside the umbrella of protection afforded to other forms of speech by the First Amendment.

The nineteenth-century movement to classify obscenity as not fit for public consumption,[17] has been followed by the recognition that interest in sexual material cannot be squelched,[18] and regulation of language must be tailored narrowly.[19]

Such a recognition makes the case that we need to better understand the function, meaning, and value of such speech in our society. Rather than bludgeoning the speech in a reflex of our own moral codes, or even as a pragmatic matter of our own social or economic survival, the higher ground might be found outside a strict legal reading of the broadcast indecency issue. We turn next to mass communication theory as a vehicle to drive us to that better understanding.

Notes

1. William K. Layman, "NOTE: Violent Pornography and the Obscenity Doctrine: The Road Not Taken," 75 *Georgetown Law Journal* 1475 (April 1987).
2. Ibid., pp. 1475–1476.
3. Ibid., at note 2, citing Cotten, "Update on Pornography," *HUMANIST*, November/December 1978, p. 48.
4. Ibid., note 7, and quoting Henry Miller's *Tropic of Cancer* (1961): "When a hungry, desperate spirit appears and makes the guinea pigs squeal it is because he knows where to put the live wire of sex, because he knows that beneath the hard carapace of indifference there is concealed the ugly gash, the wound that never heals. And he puts the live wire right between the legs; he hits below the belt, scorches the very gizzards. It is no use putting on rubber gloves; all that can be coolly and intellectually handled belongs to the carapace and a man who is intent on creation always dives beneath, to the open wound, to the festering obscene horror" (p. 225).
5. Donald M. Gillmor, Jerome A. Barron, Todd F. Simon, and Herbert A. Terry, *Mass Communication Law: Cases and Comment*, 5th ed. St. Paul: West Publishing, 1990, p. 647.

6. Ibid.
7. Margaret A. Blanchard, "The American Urge to Censor: Freedom of Expression Versus the Desire to Sanitize Society—from Anthony Comstock to 2 Live Crew," 33 *William & Mary Law Review* 741 (Spring 1992).
8. Ibid., pp. 753–754.
9. Ibid., pp. 745–746. Anthony Comstock's crusades went after such titles as, *The Lustful Turk, The Lascivious London Beauty, Beautiful Creole of Havana*, and *Fanny Hill*. Riding that momentum, other critics went further, attacking Nathaniel Hawthorne's *The Scarlet Letter* (for endorsing adultery) and Walt Whitman's *Leaves of Grass* (for "references to anatomy and sexual intercourse"). The *Comstock Act* is cited as: *An Act for the Suppression of Trade in, and Circulation of, Obscene Literature and Articles of Immoral Use*, ch. 258, @ 2, 17 Stat. 598, 599 (1873).
10. Ibid.
11. Ibid.
12. Ibid., pp. 757–758. Blanchard writes, "Erastus Beadle, who headed one of the major dime novel publishing houses, took steps to safeguard his empire. In what likely was the first code of self-regulation imposed on a communication industry because of community opposition, Beadle told his authors to obey certain rules in framing their stories. 'We prohibit all things offensive to good taste, in expression or incident,' the Beadle statement to potential authors said. Further, '[w]e prohibit subjects or characters that carry an immoral taint,' and '[w]e prohibit the repetition of any occurrence, which, though true, is yet better untold.' Finally, '[w]e prohibit what cannot be read with satisfaction by every right-minded person—old and young alike.' "
13. Ibid., p. 789.
14. John R. Bittner, *Law and Regulation of Electronic Media*, 2nd ed. Englewood Cliffs, N.J.: Prentice Hall, 1994, p. 48.
15. Kent R. Middleton and Bill F. Chamberlin. *The Law of Public Communication*, 3rd ed. New York: Longman, 1994, p. 531.
16. Ibid.: "However, in 1978, in *FCC v. Pacifica Foundation*, the Supreme Court ruled that the FCC could regulate indecent broadcasts without violating the First Amendment."
17. *Ibid.*, p. 336: "The movement for an obscenity law was led by Anthony Comstock, a strict New England Congregationalist, who had created a committee at the YMCA in New York for the suppression of vice." He later became a special post office agent appointed to enforce the Comstock Act.
18. Ibid., citing Justice Brennan's observation in *Roth*: "This 'great and mysterious motive force in human life,' he said, is 'one of the vital problems of human interest and public concern.' "
19. Ibid., at 42–44. In *Cohen v. California*, the jacket slogan "Fuck the Draft" as used "was not fighting words because it produced no immediate danger of a violent physical reaction in a face-to-face confrontation." More recently, in *R.A.V. v. City of St. Paul* (1992), Justice Scalia observed that a hate crime law aimed at stopping racist, anti-semitic, or gender-bashing speech could not be justified as "special prohibitions on those speakers who express views on disfavored subjects."

Chapter 4

Mass Communicators: Gender and Theoretical Issues

The most obvious connection between the laws of broadcast indecency and mass communication research has already been men-
tioned—that is, the scientific attempts to link media content to
antisocial effects on children or underdeveloped adults.[1] Beyond
the concern over effects, we want to consider two other important
mass communication research areas: (1) the sociological study of
mass communicators, and (2) the contemporary concern over mass
media messages as a promoter of gender stereotypes.

*Despite the common assumption by members of the general public, the
FCC and the courts that we need to regulate broadcast indecency to "protect children," an exhaustive study by Donnerstein et al. of empirical evidence failed to link indecent broadcast content to harmful effects.*[2] No
studies on the scientific effects on exposure to indecent language
were found, but the literature does suggest that younger children
have lower comprehension than older ones:

> This finding of sexual naivete on the part of younger children is
> further reflected in studies of children's knowledge of clinical sex-
> ual terms. . . . Hyde (1990), a noted sexual expert, asserted that
> American children between the ages of 5 and 15 are 'sexual illiter-
> ates. . . .'[3]

Even teenagers appear not to "hear or understand" content of
objectionable music lyrics.[4] The social science evidence, then, sug-

gests that the group adults are most concerned about—very young children—do not have the cognitive ability to decode the most troublesome content. "There is serious reason to doubt," the researchers conclude, "that exposure to such material has an effect on children up to age 12 in view of the general sexual illiteracy of this age group, their limited ability to understand sexual references, and their probable lack of interest in indecent material."[5] They go on, "Although adolescents 13–17 years old may understand indecent material, they are likely to have developed moral standards which, like adults, enable them to deal with broadcast content more critically."[6]

A Framework: Understanding Societal Fears

Why, then, do so many parents fear that America's children are being harmed? One answer might be found in the so-called third person effect—the idea that people generally worry that others, weaker than ourselves, are susceptible to harm from mass media.[7] Presumably, we are sophisticated enough to ward off the negative effects of media, but children and some adults, we might assume, may not.[8] While the basic contention of the third person effect may not be true, the belief in it may drive poor regulatory policy. In a similar sense, psychological literature tells us in the attribution theory that we link our own concerns to others—whether or not they actually have them.[9] When we argue there are societal norms of decency, we incorrectly assume that *all* children have been exposed to similar parental messages with regard to morality.

Applied Sociology

In a sociological sense, what we are talking about might be framed by symbolic interactionist Herbert Blumer as parents "seeing" in the content of broadcast indecency what is inside them.[10] That is, children do not see indecency until it is *defined for them* by outside social forces.[11] As George Carlin clearly observed in his ironic monologue, the word "fuck" may or may not be a "dirty" word because its meaning can be beautiful or ugly. In fact, the word has no meaning until our society assigns one or more meanings.

The voices in the mass media are more than distant communicators. They assist the society in creating what Craig Calhoun, a social theorist, labels "imagined communities."[12]

> That is, people have come increasingly to conceive of themselves as members of very large collectivities linked primarily by common identities but minimally by networks of directly interpersonal relationships—nations, races, genders, Republicans, Muslims, and "civilized people."[13]

In this sense, a radio star such as Howard Stern or Steve Dahl creates an *imagined community* of listeners—loyal followers, if you will. Further, these so-called shock jocks communicate through the mass media and their professional organizations and publications to create separate imagined communities. The technology, then, disregards physical or social distance to produce what communication scholars Horton and Wohl in the 1950s labeled "parasocial interaction"—where audience members come to identify strongly with mass media stars. In extreme cases, a heavy user of mass media might become isolated from real interpersonal interaction.[14]

As sociologist Erving Goffman argued many years ago, everyday human interaction is a complicated and difficult negotiation of individual perceptions:

> When we allow that the individual projects a definition of the situation when he appears before others, we must also see that others, however passive their role may seem to be, will themselves effectively project a definition of the situation by virtue of any lines of action they initiate to him.[15]

For the radio announcer and listeners, however, the communication is primarily one-way. Only a relatively few "callers" interact with a host—and then only in very controlled ways. So, the listener comes expecting to hear something. From this perspective, if the listener is not gratified in some way by the experience, he or she is likely to go elsewhere.[16]

> A bargain is involved. Sometimes, it is true, the manipulator is able to lead his audience into a bad bargain by emphasizing one need at

the expense of another or by representing a change in the significant environment as greater than it actually has been. But audiences, too, can drive a hard bargain. Many communicators who have been widely disregarded or misunderstood know that at their cost.[17]

Part of our concern with broadcast indecency, then, might be that adult broadcasters have an ability to strike an unfair bargain with unsupervised children—one in which their parents play no part in the negotiation.

Mass Communication Theory as a Framework: Defining Social Reality

Most communication scholars assume that mass media are an increasingly important part of the modern world. Theorist Denis McQuail (1994) sees media as "a major source of definitions and images of social reality; thus also the place where the changing culture and the values of societies and groups are constructed, stored and most visibly expressed."[18] He finds it difficult to define the term *mass communication*:

> The general implication from these remarks is that mass communication was, from the beginning, more of an idea than a reality. The term stands for a condition and a process which is theoretically possible but rarely found in any pure form. It is an example of what the sociologist Max Weber called an 'ideal type'—a concept which accentuates key elements of an empirically occurring reality. Where it does seem to occur, it turns out to be less massive, and less technologically determined, than appears on the surface.[19]

Mass communicators are employed by organizations working with the mass media institution. Broadcasting is seen as a distinctive media institution because of a "high degree of regulation, control or licensing by public authority—initially out of technical necessity, [but] later from a mixture of democratic choice, state self-interest, economic convenience, and sheer institutional custom."[20] Broadcast indecency, in this view, produces a normative control on content for cultural or moral reasons.[21] The legal system has a reaction to broadcasters who do not share the political definition of the way speech ought to be presented to be in the "public interest, convenience, and necessity," so defined.

While commercial success is demanded in an economic and legal sense, too much commercial success can be seen as exploitation or even a power grab by those elected or appointed to wield political control. It is clear that members of Congress, for example, see their role as in part to protect the public from social dangers. While mass communication theory is by its nature abstract, we can use it generally in the study of legal or regulatory issues.

Use of Mass Communication in the Study of Law

The use of mass communication in the study of law requires us to integrate normative legal frameworks into a social context. In short, we attempt to place legal constructions within the social context that they emerge from and exist in—law and policy are seen as serving social functions.

Consider the concept of "public interest." It is central to the meaning of the Communications Act of 1934. The allocation of broadcast facilities is based on the FCC's interpretation of the statutory language, "if public convenience, interest, or necessity will be served."[22] To the broadcast manager, however, "public interest" moves from being an abstract legal construction to an operationalized station behavioral objective. Jim Oetken writes: "At our television station, the ABC affiliate in Louisville, we have 105 people to help us serve 'inside the public interest.' "[23]

> Our management team consists of a small handful of people who set direction and policy. How effectively we serve our community is based largely on: (1) how well we choose, train, and motivate the rest of our employees, and (2) how well we communicate what we're trying to do and define their part of the plan.[24]

As both a legal and social matter, the First Amendment is designed to allow "editorial autonomy" to be exercised by media practitioners.[25] In the view of one legal scholar (Wright 1990), however, the sheer complexity of free speech law restricts a marriage of legal theory and First Amendment practice.[26]

Particularly when speaking about government regulation of mass media content, media practitioners and the general public may be confused by legal tests and constructions. Broadcast indecency challenges our belief in a near-absolute First Amendment because media practitioners may see the content as not worth the trouble, and the public may see it as socially undesirable.

Gender Studies of Language and Power

More will be said later about the application of gender studies on the question of broadcast indecency, but here we want to note that much of what passes as "broadcast indecency," as well as significant chunks of broadcast programming and commercial matter, places women (as a less powerful social group) in stereotypical negative portrayals.[27]

Predictive Tools in the Study of Indecency: Larger Concerns

Social theory may assist us in understanding the legal construction of broadcast indecency as a phenomenon that places the policy in a cultural context. The study of broadcast indecency is part of a larger concern leading to laws against public nudity and the like. In the words of Lord Delvin:

> If we thought that unrestricted indulgence in the sexual passions was as good a way of life as any other for those who liked it, we should find nothing indecent in practice of it either in public or in private. It would become no more indecent than kissing in public. Decency as an objective depends on the belief in continence as a virtue which requires sexual activity to be kept in with prescribed bounds."[28]

Conway (1977) has identified the core libertarian "dilemma" of either admitting morality can be regulated or allowing for an anything-goes position; but even British free speech philosopher John Stuart Mill would not go that far:

> There are many acts which, being directly injurious only to the agents themselves, ought not be legally interdicted, but which if done publicly, are a violation of good manners and, coming thus within the category of offences against others may be rightly prohibited. Of this kind are offences against decency; on which it is unnecessary to dwell. . . .[29]

The problem with movement away from a libertarian ideal that even Mill did not support is that it is far too easy for a society to label that which is objectionable for a variety of political and social reasons as "indecent." Such a label would seem to make it easier to find public support for narrowly based content regulation of the mass media.

Notes

1. Shearon A. Lowery and Melvin L. DeFleur, *Milestones in Mass Communication Research*, 2nd ed. New York: Longman, 1988, pp. 31–78.
2. Edward Donnerstein, Barbara Wilson, and Daniel Linz, "Standpoint: On the Regulation of Broadcast Indecency to Protect Children," *Journal of Broadcasting & Electronic Media* 36(1):111–117 (Winter 1992).
3. Ibid., p. 112.
4. Ibid., p. 113.
5. Ibid., p. 116.
6. Ibid.
7. Such an effect presumes that we are concerned about others in society. The theory, grounded in the sociology of Davison, might help explain why morality campaigns get so much public attention.
8. Consider, for example, the FCC's current position that attempts to insulate children by channeling broadcast indecency to late-night hours when they may not be present in the audience.
9. Here, we might attribute our own notions of morality to others and assume all parents want what we wish for our own children.
10. See Herbert Blumer, *Industrialization As an Agent of Social Change, a Critical Analysis*. New York: Aldine de Gruyter, 1990. Likewise, scholars "declare that industrialization introduces an arrangement of life that is strange, unfamiliar, and disturbing to people in a preindustrial society" (p. 105). I am indebted to Alexandra Shepherd Lipschultz, my spouse, for illuminating this point. The question broadcast indecency poses, then, is whether mass communicators force a disturbing culture on our society, or whether forces in the audience demand and popularize it.
11. Ibid.
12. Craig Calhoun, "Indirect relationships and imagined communities: Large-scale social integration and the transformation of everyday life," in *Social Theory for a Changing Society*, Pierre Boudieu and James S. Coleman, eds. Boulder: Westview Press, 1991, pp. 95–130. Calhoun borrows the phrase from Benedict Anderson (1983).
13. Ibid., pp. 95–96.
14. Mass media may be seen as a substitute for interpersonal human communication, which should be more two-way than the generally one-way messages broadcasters disseminate.
15. Erving Goffman, "Presentation of Self in Everyday Life," in *Introducing Sociology, a Collection of Readings*, Richard T. Schaefer and Robert P. Lamm, eds. New York: McGraw-Hill, 1987, pp. 13–18.

16. See Werner J. Severin and James W. Tankard, Jr., *Communication Theories: Origins, Methods, and Uses in the Mass Media*, 3rd ed. New York: Longman, 1992, pp. 269–281.
17. Ibid., p. 269 (quoting Davison, 1959, p. 360).
18. Denis McQuail, *Mass Communication Theory, an Introduction*, 3rd ed. London: Sage, 1994, p. 1.
19. Ibid., p. 11.
20. Ibid., p. 18.
21. Ibid., pp. 24–25.
22. 47 U.S.C.A., Section 307.
23. Jim Oetken, "Public Interest: It Starts from the Inside," in *Public Interest and the Business of Broadcasting: The Broadcast Industry Looks at Itself*, Jon T. Powell and Wally Gair, eds. New York: Quorum Books, 1988, p. 144.
24. Ibid.
25. Dom Caristi, *Expanding Free Expression in the Marketplace, Broadcasting and the Public Forum*. New York: Quorum Books, 1992, pp. 10–12.
26. R. George Wright, *The Future of Free Speech Law*. New York: Quorum Books, 1990, pp. 227–241.
27. McQuail (1994), pp. 202–203.
28. Quoting, Lord Patrick Delvin, *The Enforcement of Morals*. London: Oxford University Press, 1965, p. 120, David A. Conway, "Law, Liberty and I," in *Philosophical Issues in Law: Cases and Materials*, Kenneth Kipnis, ed. Englewood Cliffs, NJ: Prentice-Hall, 1977, p. 87.
29. Conway, pp. 877–888, quoting John Stuart Mill, *On Liberty*. New York: Liberal Arts Press, 1956, p. 119.

Chapter 5

A Content Analysis of Nonactionable Broadcasts

A proposed total ban on broadcast indecency promoted by the Congress,[1] which was adopted by the FCC but rejected by an appeals court,[2] left the U.S. Supreme Court holding its ground on the *Pacifica* decision and its progeny: "repeated use, for shock value" of "indecent" language is permissible, but only during late night and overnight hours.[3]

This left shock jocks and their stations to force a direct challenge of fines levied for language used in the key morning and afternoon radio drive times.[4] Infinity Broadcasting vowed to fight fines against the *Howard Stern Show*, but the huge corporation eventually agreed to a 1995 $1.7 million settlement in order to clear the record and the way for $275 million in station purchases.[5]

The controversy over the protection of children from indecency on the airwaves is not new. A divided U.S. Supreme Court in *Pacifica* upheld the right of the Federal Communications Commission to police some content at a time when audience ratings drove the popularity of so-called topless radio formats during midday hours.[6]

The attempts by the FCC to enforce standards today seem particularly intriguing in light of the more general "hands-off" policy of self-regulation. The regulatory record in the indecency cases is exceptional in the FCC's attempt to tell broadcasters how to do their work. One legal commentator suggested that commissioners err by assuming the role of "ultimate decision-makers" rather than turning the job over to an "objective third party" for review: "This procedure would allow for the elimination of political, personal, and other biases in the decision-making process of the FCC and give due credit to the standards espoused by the community—that is, the listeners."[7] While this view of objectivity seems simplistic, the record does reveal an extremely subjective process.

Regulating Indecent Content and Offensive Language

The debate over the regulation of radio and television content has advanced with little discussion of the offending content beyond the terms of the legal definition: "language that describes, in terms patently offensive as measured by contemporary community standards for the broadcast medium, sexual or excretory activities and organs."[8] Additionally, broadcasters have been reminded: "Programming that purely panders to prurient or morbid interests should be avoided."[9] Commissioner James H. Quello, at age seventy-six, told broadcasters that as an Army veteran, "If it offends me, it has to be pretty bad!"[10]

The so-called channeling approach now provides broadcasters with a "safe harbor" (10 P.M. to 6 A.M.) in which they are free to use the language they otherwise are not. The FCC's Order came following two United States Court of Appeals decisions in the Summer of 1995 that upheld regulatory rights:

Federal Register
Vol. 60, No. 166
Rules and Regulations
FEDERAL COMMUNICATIONS COMMISSION (FCC)
47 CFR Part 73
[GC Docket No. 92-223; FCC 95-346]
Broadcast Indecency
60 FR 44439

DATE: Monday, August 28, 1995

ACTION: Final rule.

SUMMARY: The Commission is amending its rules on enforcement of prohibitions against broadcast indecency so as to be in compliance with the instructions given by the United States Court of Appeals for the D.C. Circuit in *Action for Children's Television v. FCC.* The intended effect of the Court's instruction is to make the time periods during which the indecency ban applies the same for both public broadcasters and commercial broadcasters.

Memorandum Opinion and Order

Adopted: August 7, 1995.

Released: August 18, 1995.

By the Commission:

1. By this Order, the Commission conforms its rules to comply with the instructions given by the United States Court of Appeals for the District of Columbia Circuit in *Action for Children's Television v. FCC,* No. 92-1092 (decided en banc June 30, 1995; mandate issued July 12, 1995). Although the Court generally upheld the Commission's implementation of Section 16(a) of the *Public Telecommunications Act* of 1992, Pub. L. No. 102-356, 106 Stat. 949 (1992), relating to the prohibition on indecent programming by broadcast stations, it remanded the case to the Commission "with instructions to limit its ban on the broadcasting of indecent programs to the period from 6:00 A.M. to 10:00 P.M." *Id., slip op.* at 30. The effect of the Court's instruction is to make the time periods during which the indecency ban applies the same for both public broadcasters and commercial broadcasters. Thus, we are hereby amending Section 73.3999 of the Commission's Rules, 47 CFR. @ 73.3999, to provide that no licensee of a radio or television broadcast station shall broadcast on any day between 6 A.M. and 10 P.M. any material which is indecent.
2. Accordingly, it is ordered, that Section 73.3999 of the Commission's Rules, 47 CFR @ 73.3999, is amended as set forth below.

Appendix—Amendatory Text

Part 73, Chapter I of Title 47 of the Code of Federal Regulations is amended as follows:

Part 73—Radio Broadcast Services

1. The authority citation for Part 73 continues to read as follows: Authority: 47 U.S.C. 154, 303, 334.

2. Section 73.3999 is revised to read as follows: @ 73.3999—
   Enforcement of 18 U.S.C. 1464 (restrictions on the transmission
   of obscene and indecent material).
   (a) No licensee of a radio or television broadcast station shall
       broadcast any material which is obscene.
   (b) No licensee of a radio or television broadcast station shall
       broadcast on any day between 6 A.M. and 10 P.M. any mate-
       rial which is indecent.

Such regulatory attempts suffer from a major weakness: the FCC *assumes* that it is possible to distinguish "indecent" language from other forms. The lack of precision in defining what is and what is not indecent content under the terms of the regulatory authority, raises the issue of whether there is an empirical difference between actionable and nonactionable content. By law, when the FCC receives a complaint about an indecent broadcast, it must first decide if the content in question is actionable under the rules.

Qualitative examples of actionable content illustrate some of what commissioners find to be "patently offensive" or "pandering," but these rulings may do more to chill broadcasters than enlighten them on what is acceptable; the regulatory road from *Pacifica* to *ACT IV* (the latest court decision) was paved with repeated rejections of vagueness challenges to the FCC's definition of indecency.[11]

The purpose of this chapter is to systematically study a set of nonactionable complaints to describe the commonalties of attributes. Specifically, the goal of this study is to begin to locate the line drawn by the recent FCC between broadcast content which is determined to be indecent or not.

While review of indecent examples illustrates when a broadcaster has crossed the line, review of the nonactionable complaints should suggest how far one may go and still avert an FCC fine under the present review. The literature base for the present study includes work on gender and humor on grounds that "indecency" is best understood in context. Use of explicit language or sexual innuendo is seen as purposeful communication having to do with more than just the problematic words.[12]

A Basis for Study

The literature base for this study is found in three separate but workable areas: the normative regulatory theory of broadcast indecency, the bridge of literature on social research on communication law, and the communication work on gender and humor. The author accepts the view that the legal conflict of indecency complaints stems from social issues, and social theory is helpful in understanding the regulatory manifestations.[13]

Normative Regulatory Theory and Legal Rules

As a strictly legal policy issue, the U.S. Federal Communications Commission had been instructed by the Congress to enforce statutory language prohibiting indecent broadcasts around the clock, but the courts rejected that mandate.

One legal analysis outlined several questions that help to explain why the indecency issue continues to bubble: the lack of a clear definition for "patently offensive material," the FCC's adoption of a "standard for 'indecency' more stringent than the one applied in the field of 'obscenity,' " the use of "contemporary community standards" applied by commissioners; the "prior restraint" issues of reviews outside the license renewal process, lack of consistent FCC rules, and the FCC's own flip-flop on the channeling issue.[14] The courts have not helped matters. Consider these words from *Pacifica*: "Words that are commonplace in one setting are shocking in another . . . one occasion's lyric is another's vulgarity."

The FCC's use of a "general legal definition" came despite the fact that "there was little articulation of the rules employed to make such decisions." Further, the FCC—as the author of the present book has previously written—"has shown no interest in employing systematic methods of content analysis that might serve to define what it is about content that is objectionable."

The research community appears to be in agreement that the vague concept of "broadcast indecency" is, in effect, a social construction by policy makers and regulators. In the words of media law professor Paul Driscoll: "Most important is the need to clarify the type of subject matter which constitutes indecent broadcast material. This could be done by giving some general, yet contextual-based examples of what might or might not violate the statute."

The qualitative examples of actionable broadcast content are many and varied. Examples have been provided that identify the range of material, including:

1. Chicago announcer Steve Dahl's comment on a *Penthouse* magazine picture of Vanessa Williams: "She was licking that other woman's vagina."

2. An Indianapolis broadcast of a fake commercial: "With Butch beer, you've got a beer that goes down easy. Taste it and you'll know why it's our personal best."

3. The Prince song "Erotic City" that includes the lyrics: "If we cannot make babies maybe we can make some time . . . fuck so pretty you and me, erotic city come alive."

4. Howard Stern's airing of this comment: "When we come back from a commercial, we have a young man who wants to play the piano with his, uh, wiener. . . ."

5. The Arkansas broadcast of a telephone argument between a station manager and his son: "To all you listeners in Paris, Arkansas, don't bend over in front of my dad. Gene Williams will fuck you in the ass."

6. A Los Angeles talk show caller's description of her sexual behavior with a dog: "I'll masturbate him, and he's so big and sweet, and he really likes it."

The above examples show various settings for the content to appear including, within talk shows, songs, and banter. In some cases the reference appears to be for the sake of a joke, but in other examples it emerges from anger or to emphasize a point. Further, the content may be actionable with or without explicit use of words such as "fuck." Instead, innuendo—under the recent rulings—may be actionable.

Social Research and Communication Law

A small group of communication researchers has begun to mesh research on communication law issues with social-scientific theory. The range of study is wide, but these efforts reinforce the idea that legal research is incomplete without some analysis of the social context of the issue under review. It has been suggested that, "Law is ultimately a set of normative constraints on human behavior."[15]

Indecency regulation appears to be a legal mechanism to create behavioral boundaries of self-constraint. Therefore, vagueness may be functional or dysfunctional in promotion of free expression by creating space or zones of fear. The Federal Communications Commission uses a complaints-driven system to identify questionable content before determining whether or not the broadcast is "actionable." The system provides for the categorization of all radio broadcasts: (a) broadcasts where no complaints are filed; (b) broadcasts where complaints are filed but determined to be "nonactionable"; and (c) "actionable" complaints. The system places the burden of "street-level cop" with the public and interest groups to collect evidence and place it in the hands of the FCC for review. Thus, no matter how "indecent" a broadcast is, it is not likely to come to the attention of the regulators without some social community conflict.

Gender and Humor

Literature on gender and humor provides one base of social research for studying actionable and nonactionable content. In studies over more than thirty years, researchers find that males are more likely to attempt humor than females, and women are frequently the targets of the jokes in humor designed for male consumption.[16] Using jokes from books and comedy routines, one recent study found females 18–24 years old preferred "self-disparaging jokes" when told by male comics.[17]

Women in mass media have been shown to be portrayed as "incompetent" when the role is outside the traditional home setting, often with emphasis on emotions, attractiveness, submission and failed relationships with men.[18] Both men and women may be the targets for stereotypes employed by mass communicators, a point that raises social issues: "[T]he stereotypes that all women use sex for power or that all men feel both physical and emotional pain less deeply than women can frustrate both men and women in real-life relationships."[19] In theoretical terms, responses to jokes "depend both upon one's familiarity with the group in question as well as upon one's attitudes toward that group."[20] In one categorization of orally communicated jokes, the most frequent types were: sex, ethnic-racial, political-governmental, or about drinking, money, or family relationships. It has been argued that such humor can be socially functional as a "vehicle for expression" that cuts through defenses and tackles taboos.[21]

A study of comic postcards found sexual themes including rape, female masturbation, ongoing intercourse, excretion, male impotence, female sexual naiveté, males as sex toys, castration, male chauvinism, breasts, homosexuality, and females as sex toys.[22] Age and conservatism correlated with preferences, and men in the study had relatively high evaluations of sexist humor: "[H]umor appreciation sometimes reflects certain personality characteristics and . . . humor appreciation reveals attitudes unwittingly and/or constitutes a subtle vehicle for the intentional communication of attitudes."[23]

The symbolic reality of broadcasting may be more stereotypical than social reality, particularly for heavy media uses.[24] Most of the current research focuses on portrayal of women on television,[25] and there appears to be little recognition that radio broadcasts may function similarly. Issues of gender bias, however, have continued to surface in the study of newspaper writers.[26]

Social research on radio broadcasting continues to be quite limited. One area which has received attention is talk radio. One researcher found that talk-radio listeners may use the program as a substitute for interpersonal communication. While older people and those from "lower socioeconomic elements" were more likely to call, Joseph Turow also found lonely housewives used programs to battle "temporary loneliness."[27]

Methodology: Content Analysis

Twenty-four audio tapes of nonactionable indecency complaints were obtained through a private contractor of the Federal Communications Commission in Spring 1990. The FCC provided a limited amount of additional documentation on the cases following use of the Freedom of Information Act (FOIA).[28]

The tapes contained portions of broadcasts from ten radio stations, segments used as "evidence" by the regulators. The twenty-four tapes were divided into 171 segments represented by natural program breaks and topic shifts. Some of these breaks pre-existed on the tapes while others were researcher-imposed. A literature-based set of coding instructions was developed and pretested in Summer 1990 by seven undergraduate and two graduate communication students. The instructions were revised and coded by the author and a graduate student coder in Fall 1990.[29]

Findings

The 171 items were from ten different stations facing indecency complaints, but evidence against three of the stations accounted for 66.1 percent of the segments.[30] The average length of the coded segments was eighty-nine seconds.[31]

Despite the fact that these were nonactionable broadcasts, most of the broadcasts were identified as having occurred in either morning or afternoon drive-time shows—outside the safe harbor.

Most of the items came within two form types: within a scripted joke or within a radio talk show,[32] and over half of the broadcasts involved two speakers.[33] Over half of the broadcasts were found to contain elements of humor,[34] but in only seven of the broadcasts did one or more of the Carlin words from *Pacifica* appear.[35]

The average broadcast contained the voices of one to two males, but it was most likely to have no female voice.[36] In fifty-six of the segments (32.7 percent), the coders found evidence of content which displayed stereotypical views of men and women.[37] In some of the cases male dominance (11.7 percent) or female submission (8.2 percent) were found to be present.[38] For example, a Missouri station broadcast a feature called "Max Bar-room" with a weekly letter to the "free-ride adviser," including this fake correspondence:

My name is Jessica, and I'm not just an ordinary bimbo. I'm a rich bimbo. Anyway, I just ended a painful relationship with a so-called religious man who, quite frankly, left a bad taste in my mouth. But with that experience in my rear, I mean behind me, I'd still like to meet a new man. Start a new relationship. Isn't there anyone out there who can come close to my expectations? Signed, Jessica loving Jesus . . . in Florida.

Well, Dear Jess. Maybe your hopes were just a little too high for this man. Oh, what's good for the shepherd isn't always so good for the sheep. Don't let one bad banana spoil the bunch. Get out there. Meet some new fellas. If they don't come to you, go get 'em. Remember, relationships take a lotta hard work. So, if the last one left a bad taste in your mouth, take a different approach. And this time don't . . . blow it.[39]

In over half of the broadcasts (61.4 percent) there were sexual references in the nonactionable content, although it was most common
to find no references (38.6 percent) followed by just one reference
(25.1 percent), two (10.5 percent), or three (7.6 percent).[40] In one
case, a song, there were thirty-three separate sexual references
coded.[41] The number of sexual references was positively correlated
with the length of the segment and the number of references to
parts of the human body.[42]

The percent of the 171 broadcast segments containing various
references was measured and ranked. The presence of sexual reference was the most common attribute across segments (61.4 percent).
This was followed by segments containing references to the human
body (36.8 percent), to politics (27.5 percent), to dating, to spouses,
and to physical force (all 21.6 percent of broadcast segments).

The political references occurred in 10 scripted-joke segments,
28 talk-show segments, and 9 other segments. For example, former
U.S. Senator John Tower was the target of jokes on a Texas radio
station:

> Did you read about John Tower this morning? It's incredible. This
> guy apparently goes to peace talks and can't keep peace in his own
> home because his wife kicked him out during the '86 peace talks . . .
> for keeping a mistress and for getting a little rowdy when he had too
> much to drink. Here are some ways that John Tower says you can
> fail a field sobriety test: tell the officer you're asthmatic, and if you
> breathe you might pass out and die, and then you'll have to sue the
> police department for millions in damages, and he'd be thrown off
> the force to raise his kids on his wife's meager earnings as a cocktail
> waitress; ask the officer to hold your beer while you operate a
> breathalyzer; and the best way is—this is what John Tower says—
> offer to give the policeman a urine speciman right there since you
> have to go like a race horse anyway.[43]

In New York, meantime, a talk-show host used politics as a subject
to stimulate caller interest:

> **Caller:** Yea, but (Mayor) Koch doesn't go out with women.
> Everybody knows that.
> **Host:** Well, you talk about . . .
> **Caller:** Donald Trump says that he's got the goods on Koch,
> and Koch will not finish his term.
> **Host:** Is that right? Donald Trump. Well, Trump said Koch, ah
> Fat Eddie, is a flop.

Caller: He's got the good ones.
Host: No, he said he's a jerk.[44]

Most of the 171 segments came from two types of radio stations: talk radio formats (45.6 percent) and rock music formats (43.3 percent). Sex and politics were the dominant themes of the segments under review by the Federal Communications Commission.

The largest group, however, were the 53 segments (31.0 percent) that appeared to have no *identifiable* dominant theme.

Male and female interpersonal interaction was also measured to look for control of the communication. In the segments where women did appear, they often were telephone callers being ques-

TABLE 5.1 Ranking of Frequencies for Descriptive Attributes of Nonactionable Broadcast Indecency Content

Item Description	Number of Segments	Percent of Total Segments (N = 171)
1. Sexual references (non specific)	105	61.4%
2. Human body parts	63	36.8
3. Political references	47	27.5
4. Dating	37	21.6
5. Mentions of spouses	37	21.6
6. Violence/physical force	36	21.1
7. Children	30	17.5
8. Racial references	28	16.4
9. Religious references	26	15.2
10. Alcoholic drinking	22	12.9
11. Illegal drug use	14	8.2
12. Human excrement	14	8.2
13. Homosexuality	14	8.2
14. Oral sexual behavior	11	6.4
15. Rape	9	5.3
16. Male physical dimensions	7	4.1
17. Sex involving children	7	4.1
18. Sex involving animals	5	2.9
19. Human-animal sexual contact	5	2.9
20. Regulation of broadcasting	5	2.9
21. Female physical dimensions	4	2.3
22. Animal excrement	3	1.8
23. Bondage	3	1.8
24. Station supervision	3	1.8
25. Prescription drug use	0	0.0

TABLE 5.2   Dominant Themes of Nonactionable Broadcast
Indecency Segments

General Theme	Number of Segments	Percent
Sex	52	30.5
Politics	38	22.2
Dating	16	19.4
Religion	12	19.4
Other**	53	31.0
Total	*171*	*100.0*

**"Other" reference included segment themes relating to use of illegal drugs, as well
as a number of segments that appeared to have *no* dominant theme.

tioned by male announcers. However, those women who did call in
to stations, appeared to attempt to exercise control through conver-
sational interruptions.

These data suggest that females, by being telephone guests on
these shows, may attempt to mediate the messages of radio hosts.

In general, very little of what was found in these tapes could be
classified as "serious" dialogue. Politics was often a target for
humor; scripted jokes mirrored adolescent locker-room humor;
and discussion of serious issues (for example, banning the film *The
Last Temptation of Christ* and community tolerance for pornography)—where it did occur—appeared to shed very little light on the
topic. In Florida, one talk-show host moved the conversation away
from the central censorship position being promoted by a guest
wanting to ban the sale of adult magazines:

TABLE 5.3   Frequency Rating of Communication Behavior in
Nonactionable Broadcast Indeceny segments

Communication Behavior	Number	Percent
1. A male interrupted a female	46	26.9
2. A male asked another male a question	42	24.6
3. A female interrupted a male	34	19.9
4. A male asked a female a question	31	18.1
5. A female asked a male a question	21	12.3
6. A female asked another female a question	1	0.6

Guest: It's called "Chester the Molester." That's the one in *Hustler* magazine.

Host: Well, could you give me an example of what might be depicted in one of those cartoons?

Guest: An adult in bed with three or four children. I don't need to get graphic or anything, but . . .

Host: I think if you're going to make an accusation, you do need to get graphic.

Guest: I don't think that it's necessary. You understand what the whole idea is.

Host: No, I don't understand what the entire idea is. There could be all kinds of reasons for it. Could you get more specific, please?

Guest: Okay, they were involved in sexual contact. In bed. *Playboy* had this particular cartoon, it was described in a . . . journal . . . where a little girl was walking out the front door. A man was standing in the door, partially naked. And, the little girl turned around and—I believe she was a Brownie, Girl Scout, or whatever, selling brownies—she turned around and said: 'You call that molesting?' These are the overtones that these magazines give. . . .

Host: So you have seen these cartoons, personally, yourself. And, of course, they immediately made you have lewd and lascivious thoughts about molesting children.

Guest: Not me.[45]

In this dialogue, which was found to be nonactionable by the FCC, we can see that there may be a very fine line between "serious" and "nonserious" political discussions.

Another station had a daily feature called the "Joke of the Day," in which listeners mailed in jokes to be read on the air.[46] Several of the jokes recorded by the complainant[47] had sexual overtones:

These guys were sitting in their favorite watering hole discussing their recent dates and trying to determine the girls' professions. I guess they didn't have much chance to get to know each other before, if you know what I mean? So the first guy said, "Well my date must have been a nurse 'cause she said, 'Just lie back and relax; this won't hurt a bit.'" The second said, "Well, my date said, 'You do this over and over 'till you get it right,' so she must have been . . . what . . . a schoolteacher." Exactly. Well, the third guy said, "My date must have been a stewardess 'cause she said, 'Put this over your nose and mouth and breathe normally.' "[48]

The nonactionable segments—viewed from a content perspective—seem difficult to distinguish from actionable material, especially when one considers that eleven of the programs appeared to discuss oral sexual behavior. The FCC order upholding a $6,000 fine against a Chicago station read: "Each of the passages describes sexual or excretory activities or organs, specifically oral-genital contact, sexual activity with a child, and anal intercourse."[49]

The description of the nonactionable content raises legal issues about judicial rejection of vagueness challenges.[50] Further, review of the content from a gender perspective offers new conceptual approaches.

Discussion: Legal False Precision

The nonactionable broadcast indecency content studied here contained mostly references to and about sexual behavior. This suggests that legal constructions that attempt to distinguish actionable from nonactionable content need to be precise. The unwillingness of the Federal Communications Commission to grapple with issues of content definition when their regulation is of content raises serious questions about the reliability and validity of the procedure.

The nonactionable content found here often placed women in a negative light through sexist humor. Further, male announcers were likely to control the broadcast conversations. Where women exercised control, men were always present; and in only one of the 171 segments were two women involved in direct communication with each other. What is not known—but what needs to be—is the extent to which male dominance is pervasive throughout radio broadcasting. We know that in terms of employment, particularly of announcers and managers, there is a long history,[51] but mass communicator demographics have not been traced to influence on content.

In other words, not only is the line between actionable and nonactionable content a blurry one, it is also possible that the line between complaints and all of radio broadcasting is artificial. If so, this would raise fundamental questions about the ability of broadcast regulators to judge any content against sets of legal rules.

The findings in the present chapter support the idea that administrative review of speech character can be dangerously arbitrary. As one commentator has written, "It is those dangers with which the First Amendment is most correctly concerned and which should concern us most in evaluating the constitutionality of broadcast regulation."[52]

It is clear from the descriptions of the nonactionable data that regulatory vagueness is an important consideration in the study of broadcast indecency. The courts, rather than requiring precision, have sought the legal safety of the precedential value of *Pacifica* and its progeny. Lacking an articulated definition of indecent content, broadcasters could look to nonactionable content as one guidepost; however, they must be able to estimate community response and the subsequent FCC interpretations. Under these rules, it should not be surprising that only a handful of broadcasters nationwide would test the boundaries of acceptable speech.

Manager's Summary

Broadcast managers should be sensitized to gender issues in their programming. In most cases, content that is degrading to women will be found nonactionable, so it is an ethical rather than legal issue for managers.

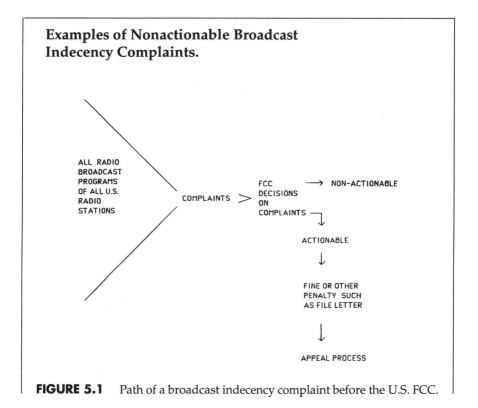

Examples of Nonactionable Broadcast Indecency Complaints.

FIGURE 5.1   Path of a broadcast indecency complaint before the U.S. FCC.

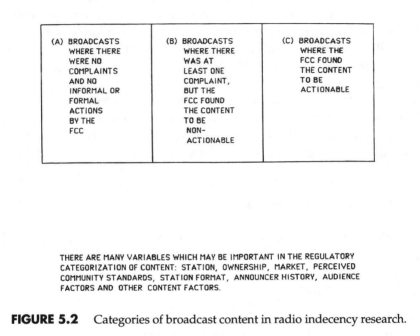

FIGURE 5.2 Categories of broadcast content in radio indecency research.

Notes

1. *Broadcasting,* "Indecency ban comes under fire, appeals court judge criticizes FCC's 24-hour indecency prohibition, challenging commission's contention that it is 'narrowly tailored,'" Feb. 4, 1991, pp. 40–41. Cf. the test in *United States v. O'Brien,* 391 U.S. 367 (1968). Justice Warren wrote that "a government regulation is sufficiently justified if it is within the constitutional power of the government; if it furthers an important or substantial governmental interest; if the governmental interest is unrelated to the suppression of free expression; and if the incidental restriction restriction on alleged First Amendment freedom is no greater than is the furtherance of that interest." A twenty-four-hour ban on broadcast indecency content—as a strictly legal issue—appears to violate at least the last two prongs of the four-part test: the interest of protecting children may be seen as related to suppression of free expression, and a total restriction seems to go well beyond the furtherance of the interest.
2. *Action for Children's Television v. FCC (ACT II),* 932 F.2d 1504 (D.C. Cir. 1991). The court relied heavily on its earlier ruling in *Action for Children's Television v. FCC (ACT I),* 852 F.2d 1332, 1344 (D.C. Cir. 1988): "Broadcast material that is indecent but not obscene is protected by the first amendment; the FCC may regulate such material only with due respect for the high value our

Constitution places on freedom and choice in what the people hear and say."
In short, even though the Congress had directed the FCC to enforce a total ban,
the court found, "While we do not ignore Congress' apparent belief that a total
ban on broadcast indecency is constitutional, it is ultimately the judiciary's
task, particularly in the First Amendment context, to decide whether Congress
has violated the Constitution."

3. See *ACT II*, 934 F.2d, at 1506, citing *Federal Communications Commission v.
Pacifica Foundation*, 438 U.S. 726, 738–741 (1978), and *In re Infinity Broadcasting
Corp. of Pennsylvania (Reconsideration Order)*, 3 FCC Rcd 930 (1987). The appellate decisions in *ACT I* and *ACT II* rely on the assumption by the *Pacifica* majority that children need protecting from indecency, and that channeling
minimizes risks.

   However, one recent conclusion is that "relevant empirical research
provides to reasonable evidence to suggest harmful effects result from
exposure to such content." See Edward Donnerstein, Barbara Wilson, and
Daniel Linz, "On the Regulation of Broadcast Indecency to Protect Children,"
Journal of Broadcasting & Electronic Media 36(1):111–117 (Winter 1992).

   Other social research on indecency regulation includes: Theodore L.
Glasser, "The Press, Privacy and Community Mores," unpublished paper
prepared for presentation to the Mass Communication Division, Speech
Communication Association, Louisville, November 1982; Mike Hindmarsh,
" 'How Is Pornographic?' (Not 'What Is Pornography?')," unpublished paper
prepared for presentation to the Communication Theory and Methodology
Division, Association for Education in Journalism and Mass Communication,
Minneapolis, August 1990; Howard Kleiman, "Indecent Programming on
Cable Television: Legal and Social Dimensions," *Journal of Broadcasting &.
Electronic Media* 30(3):275–294 (Summer 1986); Harvey Jassem and Theodore L.
Glasser, "Children, Indecency, and the Perils of Broadcasting: The 'Scared
Straight' Case," *Journalism Quarterly* 60(3):509–512 (Autumn 1983); and Gray
Cavender, " 'Scared Straight': Ideology and the Media," in *Justice and the
Media, Issues and Research*, Ray Surette, ed. Springfield, Ill.: Charles C. Thomas
(1984), pp. 246–259.

4. *Broadcasting*, "Infinity to Fight FCC Indecency Fine," 3 December 1990, p. 38.

5. "Infinity buys stations for $ 275 million," *Broadcasting & Cable*, 25 September
1995, p. 11.

6. John C. Carlin, "The Rise and Fall of Topless Radio," *Journal of Communication*
26(1):31–37 (Winter 1976); also see Guy A. Reiss, "New F.C.C. Standards on
Indecency on the Air and the First Amendment: Offensive Obscenity or
Profound Profanity," *Columbia-VLA J.L.&A.* 13:221 (1989); Mathew L. Spitzer,
*Seven Dirty Words and Six Other Stories, Controlling the Content of Print and
Broadcast*. New Haven: Yale University Press, 1986; Jeremy Harris Lipschultz,
"Conceptual Problems of Broadcast Indecency Policy and Application,"
unpublished paper prepared for presentation to the Radio-Television
Journalism Division, Association for Education in Journalism and Mass
Communication, Minneapolis, August 1990; and Paul D. Driscoll, "The Federal
Communications Commission and Broadcast Indecency," unpublished paper
prepared for presentation to the Law Division, Association for Education in
Journalism and Mass Communication, Washington, D.C., August 1989.

7. Reiss, op. cit., p. 243. Also see, Erwin G. Krasnow, Lawrence D. Longley, and Herbert A. Terry, *The Politics of Broadcast Regulation*, 3rd ed. New York: St. Martin's Press, 1982.; Haeryon Kim, "The Politics of Broadcast Deregulation: Beyond Krasnow, Longley, and Terry's 'Broadcast Policy-Making System,' " unpublished paper prepared for presentation to the Association for Education in Journalism and Mass Communication, Minneapolis, August 1990; T. Barton Carter, Marc A. Franklin, and Jay B. Wright, *The First Amendment and the Fifth Estate, Regulation of Electronic Mass Media*, 2nd ed. Westbury, N.Y.: The Foundation Press, 1989, pp. 276–345; and Michael Starr and David Atkin, "The Department of Communications: A Plan for the Abolition of the Federal Communications Commission," unpublished paper prepared for presentation to the Radio-Television Journalism Division, Association for Education in Journalism and Mass Communication, Washington, D.C., August 1989.

8. *In the Matter of Enforcement of Prohibitions Against Broadcast Obscenity and Indecency in 18 U.S.C. 1464*, Order, FCC 88–416, 19 December 1988; and 53(249) FR 52425, 28 December 1988.

9. 1990 FCC Lexis 3629.

10. *Broadcasting*, "Life as a Washington Monument," 31 December 1990, p. 55.

11. "Court Adds 'Certainty' to Indecency Policy, *Broadcasting & Cable*, 24 July 1995, p. 65. Not only has vagueness been sidestepped by the courts, the *ACT IV* court found that the lengthy process—up to seven years—had not violated broadcasters constitutional rights.

12. Karen K. Dion and Kenneth L. Dion, "Personality, Gender, and the Phenomenology of Romantic Love," in *Self, Situations and Social Behavior*, Phillip Shaver, ed. Beverly Hills: Sage, 1985, pp. 209–239.

13. Media scholar Doris Graber, for example, sees FCC content regulation as a form of "performance control." For a cultural approach to the issue, see Edward A. Shils, "Mass Society and Its Culture," in *Reader in Public Opinion and Communication*, 2nd ed., Bernard Berelson and Morris Janowitz, eds., New York: The Free Press, 1966, p. 526: "Popularization brings a better content. . . . Of course, men will remain men, their capacities to understand, create, and experience vary, and very many are probably destined to find pleasure and salvation at other and lower cultural levels."

14. Reiss, op. cit.

15. Jeremy Cohen and Timothy Gleason, *Social Research in Communication and Law: The Sage CommText Series*, Vol. 23, Newbury Park, Calif.: Sage, 1990, pp. 14, 28.

16. Gretta Kontas, "Gender, Disparaging Humor and Aggression, Have We Come Far Enough to Laugh, Baby?" unpublished paper prepared for presentation to the Speech Communication Association, November 1990; Joanne R. Cantor, "What is Funny to Whom? The Role of Gender," *Journal of Communication* 26(3):164–172 (Summer 1976); and Anthony J. Chapman and Nicholas J. Gadfield, "Is Sexual Humor Sexist?" *Journal of Communication* 26(3):141–153 (Summer 1976).

17. Kontas, op. cit., p. 5.

18. Laura A. Terlip, "Tough, Tender & Too Good to be True?: Student Attributions of Sex Roles to Successful Females in Situation Comedies," unpublished paper prepared for presentation to the Women's Caucus, Speech Communication Association, November 1990.

19. Lea P. Stewart, Alan D. Stewart, Sheryl A. Friendly, and Pamela J. Cooper, *Communication Between the Sexes, Sex Differences and Sex-Role Stereotypes*, 2nd ed. Scottsdale, Ariz.: Gorsuch Scarisbrick, 1990, pp. 208–209.
20. Jeffrey H. Goldstein, "Theoretical Notes on Humor," *Journal of Communication* 26(3):104–112 (Summer 1976), p. 106.
21. Charles Winick, "The Social Contexts of Humor," *Journal of Communication* 26(3):124–128 (Summer 1976), pp. 124, 128.
22. Chapman and Gadfield, op. cit.
23. Ibid., p. 151.
24. Tamar Zemach and Akiba A. Cohen, "Perception of Gender Equality on Television and in Social Reality," *Journal of Broadcasting & Electronic Media* 30(4):427–444 (Fall 1986).
25. Rita A. Atwood, Susan Brown Zahn, and Gail Webber, "Perceptions of the Traits of Women on Television," *Journal of Broadcasting & Electronic Media* 30(1):95s101 (Winter 1986).
26. George Comstock, *The Evolution of American Television*. Newbury Park, Calif.: Sage, 1989, pp. 182–183.
27. Joseph Turow, "Talk Show Radio as Interpersonal Communication," *Journal of Broadcasting* 18(2):171–179 (Spring 1974). See pp. 176, 178.
28. The *Freedom of Information Act* was employed (5 U.S.C. 552), FOIA Control No. 89–217, with the request renewed in a letter dated Oct. 1, 1990. The FCC delayed response to the requests on numerous occasions during a two-year period.
29. When information was forthcoming, there were limitations. For example, the FCC invoked "confidential" privacy exemptions to deny requests to identify some who complained on grounds it might inhibit future station employees from coming forward. In an October 1990 personal visit to FCC offices, some materials were copied by the author. The FCC, however, has failed to send additional materials that were promised, and another set of thirty actionable tapes has yet to be delivered.

    The stations were: 1. KWTO-FM, Springfield, Missouri (one tape); 2. KZFM-FM, Corpus Christi, Texas (twelve tapes); 3. WAAV-AM, Wilmington, North Carolina (two tapes); 4. WEBN-FM, Cincinnati, Ohio (one tape); 5. WFLA-AM, Tampa, Florida (one tape); 6. WMCA-AM, New York, New York (three tapes); 7. WSMC-FM, Collegedale, Tennessee (one tape); 8. WSPN-FM, Saratoga Springs, New York (one tape); 9. WTMA-AM, Charleston, South Carolina (one tape); and 10. WYSP-FM, Philadelphia, Pennsylvania (one tape). Additionally, KSHE-FM, Crestwood, Missouri was under review for "Dirty Boulevard," and a transcript was made by the FCC. Ole R. Holsti, *Content Analysis for the Social Sciences and Humanities*, Reading, Mass.: Addison-Wesley, 1969. Cf. Joseph R. Dominick and James E. Fletcher, *Broadcasting Research Methods*, Newton, Mass.: Allyn and Bacon, 1985, pp. 12–13. Reliability coefficients were consistently above .70: drinking, .93; racial, .88; politics, .85; religion, .93; dating, .93; show type, .84; stereotyping, .75; dominance, .90; and submission, .91.

    See Akiba A. Cohen, Hanna Adoni, and Charles R. Bantz, *Social Conflict and Television News*, Sage Library of Social Research 183, Newbury Park, Calif.: Sage, 1990, p. 69: "[W]e have chosen to join the growing trend away from doctrinal use of statistical hypothesis testing. Thus, we present the data without resorting to conventional tests of statistical significance. We regard the statistics as descriptive in nature and look upon our treatment of the information gathered in this study as exploratory.

30. The three stations were: WEBN, WMCA, and KZFM, two rock radio stations and a talk radio station.

31. The range was from 10 to 285 seconds.

32. Of the segments, 74 (43.3 percent) were coded as rock formats; 78 (45.6 percent) were from talk formats; and the remaining 19 (11.1 percent) were other or don't knows.

33. Of the segments, 89 (52 percent) had two speakers; 46 (26.9 percent) had one; 28 (16.4 percent) had three; and 8 (4.7 percent) had four to five speakers.

34. This was consistent with the literature review that suggested humor often portrays women in stereotypical ways. In 91 segments (53.2 percent) humor was found to be present.

35. A key failure of the *Pacifica* decision was that its focus on the Carlin words obscured the more difficult issue of regulating sexual innuendo.

36. Of the segments, 94 (55 percent) contained one male voice; 61 (35.7 percent) contained two males voices; but 93 of the segments (54.4 percent) contained no female voices. In 69 (40.4 percent) of the segments, one female voice was found to be present.

37. This variable, however, was the most difficult of the set to obtain intercoder agreement. Ultimately, agreement scores were above 70 percent. Researchers need to develop better ways to identify absence or presence of gender stereotypes.

38. The dominance-submission subject seems ripe for further study because it addresses "indecency" from a communication theory perspective. See Gretchen S. Barbatis, Martin R. Wong, and Gregory M. Herek, "A Struggle for Dominance: Relational Communication Patterns in Television Drama," *Communication Quarterly* 31(2):148, 155 (Spring 1983): " 'Theatrical friction' aside, one interested in the development of gender role content with the potential for prosocial effect might well ask, 'Why can't a man also be more like a woman?' " Also see Stephen B. Groce and Margaret Cooper, "Just Me and the Boys? Women in Local-Level Rock and Roll," *Gender & Society* 4(2):220–229 (June 1990).

39. KWTO-FM, Springfield, Missouri, segment number 028.

40. A generalization can be made that most often the sexual references were in passing.

41. However, the references were, again, innuendo of "slip it in" amid moans and groans of a male and female seemingly involved in sexual intercourse.

42. The Pearson product moment correlations were .48 for the number of sexual references and broadcast length, and .47 for the number of sexual references and number of references to body parts. These were statistically significant at the alpha = .05 level using the t-statistic.

43. KZFM-FM, Corpus Christi, Texas, segment number 130. Political humor was often entangled with sexual reference. This is an important problem because "pandering," by definition, would have no serious value. Indecency is often classified as "nonpolitical," or less protected speech.

44. WMCA-AM, New York, New York, segment number 059.

45. WFLA-AM, Tampa, Florida, segment number 048.

46. WEBN-FM, Cincinnati, Ohio.

47. The complaintant was Capt. Theodore J. Schoch of the Cincinnati Department of Public Safety. In a letter dated Oct. 26, 1989, Edythe Wise, chief of the Complaints and Investigations Branch, Enforcement Division, FCC Mass Media Bureau replied to complaint 8210-EJS: "[W]e conclude that the broadcast material identified in your complaint is not actionably indecent. Accordingly, while we recognize that the material in your complaint may be offensive to many, we cannot find the necessary legal basis for further Commission action."

In a transcript from September 1987 (about 7:53 A.M.) there was evidence that several jokes of the day were studied by the FCC.

48. Ibid., segment number 016.

49. *Broadcasting*, "Indecency Ban Comes under Fire," 14 February 1991, pp. 40–41. Oral sex appeared to be a target of FCC concern.

50. Apparently, even content prohibitions on oral sex as a topic for radio talk must be qualified. Tom W. Smith, "The Polls—A Report, The Sexual Revolution?" *Public Opinion Quarterly* 54:415–435 (1990); Cf. Peter Alan Block, "Modern-Day Sirens: Rock Lyrics and the First Amendment," *S. Cal. L. Rev.* 63:777 (1990); Steven G. Gey, "The Apologetics of Suppression: The Regulation of Pornography as Act and Idea," *Michigan L. Rev.* 86:1564 (1988); Heidi Skuba Maretz, "Aural Sex: Has Congress Gone Too Far by Going All the Way With Dial-a-Porn?" *Hastings Comm/Ent L.J.* 11:493 (1989); and Ronald Berman, *Advertising and Social Change.* Beverly Hills: Sage, 1981, 54: "The presumption is that advertising is determined to undermine the sexual revolution. It neutralizes feminism by depicting women with a new sense of assertion and aggressiveness, but still confined to a household environment.

It represents the female body only in order to suggest that each part must be deodorized, sprayed, or depilated. It suggests that women are infantile and to some degree autoerotic or capable of finding satisfaction principally through the embrace of commodities."

51. While women have made gains in terms of broadcast employment, men still dominate the field, particularly at the management level. See Barry L. Sherman, *Telecommunications Management, The Broadcast & Cable Industries*, New York: McGraw-Hill, 1987, pp. 116–117. Women represented 34 percent of employees, minorities 17 percent.

52. Timothy B. Dyk, Book Review, "Seven Dirty Words and Six Other Stories," 40(1) *Federal Communications Law Journal* 40(1):131, 141 (February 1988).

Chapter 6

The Role of Audience and Community in Complaints

The regulation of "indecent" broadcasts in the United States has been a topic of concern to researchers interested in the constitutional questions of law, as well as to those who are curious about social implications. For industry and government officials, the issue appears to be significant in terms of redefining future relationships. And for interest groups, the complaints lodged against some radio and television stations during the 1980s appeared to be a tool to lobby for change. Broadcast managers need to know about the role that station audiences and community members play in the legal and social arenas.

Case Studies: Contrasting Two Examples

Following a review of the history of the issue and recent developments, the present chapter focuses on two radio stations—WLUP-AM, Chicago and KSJO-FM, San Jose—to study the content in question, and the community reaction to it.

Federal Communications Commission files and tapes were accessed in a visit to the Washington, D.C., agency offices in October 1990 and by making a series of Freedom of Information Act requests during the 1988–1991 period. In this chapter, we seek to improve understanding about the nature of the role that a station's audience may play, as well as to that of the larger community in a service area covered by the broadcaster's signal.

Regulation in a Legal Context

While challenges to the vagueness of the FCC definition of indecency ("language that describes, in terms patently offensive as measured by contemporary community standards for the broadcast medium, sexual or excretory activities and organs")[1] have been rejected, the United States Court of Appeals protected the rights of broadcasters to air such material during late-night hours when children were not expected to be present in the audience. The Court interpreted the U.S. Constitution as mandating a "reasonable safe harbor rule" rather than a "blanket ban."[2]

Doubt has been raised about authority through the Communications Act of 1934 to override the broadcaster role of self-regulation—that is, knowing best what the audience wants.[3] Second, the FCC historically has reacted to complaints rather than initiating probes—a process that makes some review appear to be arbitrary.[4] Third, the decisions on complaints as being either "actionable" or "nonactionable" do not seem to be systematically based.[5] And fourth, the process and effects of regulation of indecent broadcasts seem to be isolated from the social realities of the local communities of interest in the stations and their broadcasts.[6]

Research from a Social Context

Some researchers interested in communication and law issues such as broadcast indecency, argue that it is necessary to place normative-legal concerns into the broader context of social research on mass communication. To the extent that actionable and nonactionable indecent broadcasts can be seen as mass communication content, theoretical development and research on gender issues in humor are relevant. Much of what has been presented as "shock radio" may be categorized as announcer-attempted humor, as we have seen.

The Audience

The role of audience cannot be ignored because there may be meaningful *feedback* from local audiences and community members to their favorite or disliked local radio talent. Further, some announcers encourage communication between audience and station. And beyond industry self-regulation, regulators of broadcasting must rely on audience members and the community for "data" on station performance in serving the "public interest, convenience, and necessity."[7]

In the history of indecency complaints against radio stations, the audience and community have played important roles. In the 1950s, for example, the norms for appropriate interaction between audience members and broadcasters were much more restrictive. Thus, a comment that would be found harmless today was seen by some as indecent nearly a half century ago. William Ray (1990), a former FCC employee, found that in a 1959 case against KIMN, Denver, a card from a listener who said she took KIMN radio with her wherever she went led to the following joke: "I wonder where she puts KIMN radio when she takes a bath. I may peek. Watch yourself, Charlotte!"[8] The important point, here, is that audience interaction played a role in the problem for the station. In a separate attempt at humor, the station employed a sound effect of a flushing toilet: "Say, did you hear about the guy who goosed the ghost, and got a handful of sheet?"

The landmark *Pacifica* case, in which the U.S. Supreme Court upheld the narrow regulation of profanity, was initiated by one unhappy listener. The so-called topless radio case of WGLD-FM made its way to the U.S. Court of Appeals because of only two Chicago organizations. And, the current concern over protection of children seems to rest on the ability to identify actual numbers of underaged listeners in the audience at various times of day and night.[9]

WLUP-AM and KSJO-FM

Two broadcast indecency investigations by the FCC shed light on the varying roles that audience and community can play in the life of a complaint. WLUP-AM, Chicago, and KSJO-FM, San Jose, were selected from a larger database because the cases each offered extensive records of audience involvement in providing evidence to the FCC about the complaints, and because the cases are contrasts in the favorable and unfavorable roles feedback can play for a station under review.

WLUP-AM

On November 30, 1989, the FCC sent the licensee of WLUP-AM a "Notice of Apparent Liability for a Forfeiture" following complaints about *The Steve and Gary Show* on the Chicago station.[10]

The notice attached transcripts from broadcasts at issue: 2:30–4:30 P.M., August 19, 1987, and 5:10 P.M., March 30, 1989. The station had argued that the broadcasts were not "indecent" under the rules; rather, the shows were "extemporaneous, open forum, frank, live, and spontaneous, often humorous, with over half of the dialogue supplied by listeners" who were screened.[11]

In the March 30 broadcast, Steve Dahl, Gary Meier, and Bruce Wolfe were discussing *Penthouse* magazine pictures of Vanessa Williams and a television interview with her:

> **BW:** He's (Bob Costas) trying to defend this Vanessa Williams. I mean she's a, the most embarrassing pictures that you ever saw in your life in *Penthouse*. I don't know and he's talking about how she actually had more talent than your typical Miss America. I'll say.
> **SD:** She was licking that other woman's vagina. I want to tell you pal.
> **GM:** Isn't this great. We can all break into Jack Brickhouse at any moment.
> **BW:** They chorus. But I mean, he's just offending her (unintelligible).
> **SD:** Went down on that other woman and oh God, you had your tongue in her vagina. It was fabulous. A lot of your Miss Americas can. I don't picture Phyllis George being able to do that.
> **GM:** More on later with Vanessa Williams (laughter).
> **BW:** Show all these pictures of her right? They're getting increasingly suggestive. They show in that break.
> **SD:** He had to put his leg up on the ottoman because he had a stiff-oh. God, I'm as hard as a rock right now Vanessa. You're so honest, Bob.
> **GM:** On the big chair.
> **BW:** Oh, he's always dropping these references.
> **SD:** He looks like he could (unintelligible). That chair's way too big.

The station defended the broadcast by calling it a "straightforward description" of the pictures: "[T]he broadcast comments were political and social commentary and gave a 'simple, accurate description of a picture of oral-genital contact,' in language that was explicit but not erotic or suggestive."[12] The innuendo, the station told the FCC, was not easily understood by children who might be in the audience.

In the case of the August 19, 1987, broadcast, the "Kiddy Porn" song was "sung by a call-in participant, contained only mild innuendo," the station argued, "and the called-in joke alluding to homosexual activity contained one double entendre" not "readily understood" by children:

> **C:** I want you guys to listen to my Cliff song and see if I'm on the right track, a little bit because I'm stumbling along with some lyrics, or, I have a—
> **SD:** Why don't you pick what you think is your best.
> **C:** OK, let's do this then because I want, I want to know if I'm on the right way. OK? You got to imagine Cliff singing this' cause it's not that funny otherwise, but this parody to September Morn and we're going to call it Kiddie Porn. Here we go. (music)

Welcome to my camp, you're a grown up little man,
let me rub out that cramp, my, you've got a lovely hand,
look at who you are and where you want to go,
drop your drawers and strike a pose,
my tripods' all aglow, it's Kiddie Porn.
Do you remember when we filmed that hot summer day,
you were nude and you discovered I was gay.
Yes, I know but let's go roll out in the hay.
(laughter in the background during song)

> **SD:** That's pretty good.
> **GM:** I think you should sing that. I think Cliff commissioned—
> **C:** I was there.
> **SD:** Pretty funny with you singing it.

In the station response to the complaint, the FCC noted that WLUP-AM's licensee recognized that "[t]he complainant likely objected to this song because he/she believed that it was improper to address the troubling issue of child pornography in a humorous tone."[13]

A final example transcripted by the FCC was a caller joke:

> **C:** I have a joke for you guys and think you can say it on the air. It's not real bad but do you know what the number one line in a gay bar is?
> **SD:** What?
> **C:** May I push the stool in for you?

SD: (laughs) Don't you think those guys have enough to deal with? We got AIDS and all that and then we're all naming gerbil jokes and stuff. I don't know.

On the strength of the statutory indecency prohibition (18 USC 1464), the precedent-setting *Pacifica* case, the *ACT I* case, FCC action against Infinity Broadcasting in 1987, and the U.S. Supreme Court language in *Sable Communications*, the FCC found: "We believe that all the subject broadcasts fit squarely within our definition of indecency. . . ."[14] Thus, the FCC levied fines totaling $6,000, including $2,000 for each of the three incidents—fines that licensee Evergreen Media refused to pay in December 1991: "Our [on air] talent," the station manager said, "deserves more guidelines on what is and is not acceptable."[15] Despite early hard-line stands by broadcasters, however, the 1990s resolution to these cases was settlement. Broadcasters such as Evergreen and Infinity were allowed to pay fines and expunge the record of complaints. By admitting no wrongdoing in the settlements, media corporations were allowed to renew existing licenses and be deemed qualified to obtain new ones.

Complaints and responses. The FCC action grew from eight handwritten complaint letters, one typewritten complaint letter, and eight tapes it had received from two individuals.[16] The FCC, by request of the two, agreed to treat them as confidential sources. Thus, access to the complaints was limited to typed transcripts of the letters. The earliest dated complaint letter sent to the FCC's Washington office is dated March 8, 1989:

```
March 8, 1989

Edythe Wise
F.C.C. Mass Media Bureau
Washington, D.C.

Dear Ms. Wise

I tuned in WLUP-AM in Chicago today at 4:55 p.m.
This is a show hosted by Steve Dahl. He is known in
```

```
this area as one of those Shock Jocks, and anything
goes. I was offended by his monologue which included
3 mentions of the penis of some other station per-
former, and several references to a competing
announcer's testicles. Enclosed is a tape I made. I
have complained before and nothing is ever done. Do
we have rules or not. In recent weeks this guy has
used words like: Bastard—Bitch—Asshole—fag-off etc.
The radio company defines it as humor. Is this funny
to you? How many tapes does it take? How many com-
plaints do you have to have?

Yours Truly,
[Deleted]¹⁷
```

Another letter dated March 24 accompanied a tape documenting references to the words "vagina," "pussy," "kiss ass," "cover my ass," and "give head."[18] The FCC also received a March 25 newspaper clipping that tied WLUP to "anti-semitic jokes." In an opinion piece on radio and television racism, the author quoted WLUP's operations manager as saying their programming was not "blatantly offensive" to religious groups: " 'We don't have our finger on the dump button' when a caller talks about 'Jewing down.'"[19] The person who sent the clip to the FCC wrote: "They say you won't do anything. I hope that they are wrong. Please return the airwaves to responsible hands."

Another letter dated March 31 included more taped evidence: "This tape contains references to oral sex and the D.J. talks about lewd behavior using a 'vagina.' It sickens me to hear this stuff almost every day." More evidence was forwarded June 14. "[T]hey say they can do what they want because there are no rules. I encourage you to look into this place. I read that they are up for their license to be renewed. Is that in Chicago's best interest?"

As the FCC began to take action against WLUP and the other shock jocks, a few more letters from Chicago continued to build a case against Steve and Gary:

August 25, 1989

Dear Ms. Killory:

I read in the Chicago papers about F.C.C. action
against indecent radio broadcasting by such stations as WLUP-AM here in Chicago. I am thrilled to
see some action against a very serious problem. I
have written several complaints to the F.C.C.
about this station.

In the last two or three weeks I have heard the
following verbiage on *The Steve and Gary Show*:

1. A broadcast from the Chicago Lakefront where
Steve suggested to young girls in a live public
broadcast that if they focused properly with the
way he was sitting in a lounge chair that they
might "Enjoy A View Of His Balls."

I have also heard words such as penis, vagina,
pussy, asshole, bastard, and more used in numerous
commentaries and skits. My tape recorder wasn't on
during the above moments, but I am writing to let
you know that filth continues here in Chicago.
Please pursue this. I am no prude, but I can't
accept this gross verbal abuse of the public's
airwaves. Please forgive my need to be graphic.

Sincerely,
[Deleted]

In its defense, WLUP-AM had letters on file from more than forty
satisfied listeners who said they enjoyed the programming.[20] The
station response, dated October 10, 1989, indicated that the FCC
refused to allow the station to face its accuser: "Because we remain
unaware of the identity of the complainant, this response cannot
adequately address whether the complainant is unusually sensitive
or is biased against WLUP (AM)."

The *Steve and Gary Show*, the station argued, met the needs of the Chicago audience and community:

> Because they seek to appeal to the average member of the Chicago community, Steve and Gary use the common vernacular rather than highfalutin language. If Steve and Gary succeed in addressing controversial issues, which the Commission has explicitly stated are not taboo, in a tone and manner similar to that used by the average member of the Chicago community, it is necessarily strong evidence that the material is not patently offensive by contemporary community standards—especially in the total absence of evidence to the contrary.[21]

WLUP told the FCC it has received "approximately 500 letters" that it characterized as "ringing endorsements." The listener letters were used as evidence to "demonstrate that a diverse listenership, which spans all segments of Chicago's adult community, appreciates the material broadcast. . . ."

The letters, which the station waived before the FCC, came from such surprising sources as a couple in their sixties, a director for religious education at a Roman Catholic Parish in Logan Square, an elementary schoolteacher, and mothers of small children. Because listeners said they used the program to relax to and from work, or as a "mid-day 'release.' " WLUP said "spontaneity" and "listeners ability to participate" were unique: "Those assets of the show also make it impossible to replicate the program in another medium or at another time of day."

It is clear from the evidence that Steve and Gary, faced with an indecency complaint, solicited from their listeners letters that could be used in response. In one, a parent said the program addressed important issues—such as the criticism of a utility company promotion:

> The service that Com Ed provides to school aged children that if they are in trouble they should go to a policeman or a Commonwealth Edison lineman or even a Com Ed employee in one of their tan cars. Mr. Dahl brought up the point that as a parent he did not want teachers or other groups, i.e., Com Ed, directing his children to strangers in automobiles.[22]

The listener letters focused on what was framed as an "unwarranted attack" on the program:

> Personally, I take umbrage at the fact that somebody who doesn't even know me is trying to dictate to what I may listen. Also, I am outraged that my friends (and that is what I consider them to be) are being persecuted on the whim of a tiny handful of non-fans. I understand that there exists a small segment of society that does not like their show (nobody that I know of), but that is understandable. I do not like asparagus; so I do not eat it. I do not try to prevent others from eating it. I just don't order it for myself.[23]

One of Steve and Gary's listeners for ten years mailed a handwritten note that said, "[I]f listening to Steve and Gary is truly detrimental to youth, I must be an exception. I'm 27 now, have no criminal record, am well-educated and hold a good job."

Some listener letters addressed the FCC directly. One argued that Steve and Gary's "spark of imagination in the vacuum [*sic*] of AM radio" produced diversity in the marketplace: Concerning the "Big Brother" F.C.C.—goes [*sic*] their investigation has no credence. If someone doesn't like certain words being said over the air, by all means they should change the station. I, however, enjoy frank, honest, and adult manner in which they broadcast.[24]

One mother of two, who called herself a "Christian who never misses church or Steve and Gary," said: "I do not want a few so called Christian do gooders to dictate what I listen to. I'm a mother and have no problems turning my car radio on when my children are in the car to AM 1000's 2:30 to 7:00."

The WLUP-AM case is similar to the following KSJO-FM case in that, in both, criticisms were leveled in the local newspaper. But the level of organized social criticism, as we will see, never developed in Chicago as it did in San Jose.

The fact that Steve and Gary asked their listeners to write letters of support to the station manager appeared to create or reinforce a level of support within the broadcast organization that was not found in the KSJO-FM case.

KSJO-FM

The indecency complaints against *The Perry Stone Show* on KSJO-FM San Jose appeared to stem from wider concerns about the broadcasts. The morning radio complaints boiled down to eight transcripted examples.[25] These included the voice of Perry Stone (PS), a male voice (MV), a female voice (FV) and callers (CA), as was the case in a broadcast from the 7 A.M. hour on October 25, 1988:

CA: Hello

PS: Sophie, baby.

CA: What do you want?

PS: Sophie baby, I love 'ya. You know how much I love you?

CA: I love you too, Perry.

PS: You know how much I appreciate you?

CA: What do you want?

PS: I love you. I'd love to, I'd love to lick the matzo balls right off your butt. As a matter of fact, I'd put it right there in the middle.

CA: Perry, you didn't get enough sleep last night. That's why you're talking like this.

PS: I'd like to have a smorgasbord in your butt. I'd like to have matzos in there with borsch, everything. I'd just have a buffet.

CA: You wouldn't find your way out, darling.

PS: I know that.[26]

A second transcript was from a song to the tune of *The Beverly Hillbillies* theme:

Come a listen to story about a man named Boas,
a poor politician that barely kept his winky fed,
then one day he's poking a chick
and up from his pants came a bubbling crude.
Winky oil.
Honey pot.
Jail bait.
Well, the first thing you know, old Roger's doing jigs
Some kinfolks said, "Rog, move away from there."
Jail bait is the place where I really want to be.
So he loaded up his winky and he did it with Beverly.
Big breasts.
Only 15 years old.
Going to jail.
Well, now it's time to say good-bye to Boas and his friends
and they would like to thank you folks for kindly dropping in.
You're all invited back to this jail serenity
and hope you bring some chick called Dorothy.

Another example included multiple voices:

PS: All right man. Thank you.

MV: When you were born maybe your parents were just too lazy. Now that you're all grown up and you're a healthy American male you have to suffer the consequences.

FV: Ohh, it's an uncircumcised one. Ohh, how gross.

MV: Yes, over 65 million healthy Americans suffer from uncircumcised ones. Unlike normal men, uncircumcised males have to constantly worry about pulling back their foreskins every time they have to use it. Now, you don't have to worry again because Rodgo proudly presents the foreskin garage opener. It's a computer chip that fits snugly on the tip of your thing and included is a remote control that you keep in your pocket. You use it just like a garage opener. Open it up when you have to use your thing. (Mechanical sound effect) Watch the foreskin get pulled back in a comfortable, relaxed fashion. Now, you're ready for some serious pounding. (Metal hammer sound effect) So, never worry again about foreskin. Pick up Rodgo's new foreskin garage opener. You'll be glad you did.

MV: Oh, wow. I think she's ready to do it. Hey, you ready to do it? Yeah, I think she's ready to do it. (Mechanical sound effect)

FV: Yeah. Oh, yeah, you are.

MV: Thank you, foreskin garage door opener.

Lawyers for KSJO-FM's now former licensee Narragansett Broadcasting,[27] in their September 25, 1989, response to the complaints, attempted to distance the licensee from the content. KSJO claimed it had not kept transcripts of the programs and could not comment on their accuracy. They noted for the FCC that Stone was suspended on March 14, 1989, and fired on March 21, 1989. They provided three reasons for not controlling Stone's behavior:

(1) Stone's repeated promises to conform to our guidelines and "clean up his act,"
(2) his program's strong and extremely loyal following and support among KSJO's adult listeners, and
(3) as will be discussed further below, genuine uncertainty on the part of the station management concerning the "contemporary community standards" that were applicable to Stone's program in regard to KSJO's audience.[28]

That is, the station pointed to external issues of Stone's audience and the community standard that went beyond their ability to control the content of the radio program.

It is noteworthy, however, that Stone was not fired for his sexual remarks: "The specific incident that triggered Stone's suspension involved his improper on-air comments (not sexual in nature) to two children who were guests on the program, but by this time KSJO's manager had already determined that Stone was insufficiently responsive to program direction and could not be persuaded to satisfactorily conform to KSJO guidelines.

The KSJO general manager, according to Narragansett attorneys, understood he was to "keep Stone's material within acceptable station and FCC bounds of taste and decency as to sexual matters, without unduly constricting ribald forms of humor which plainly appealed greatly to a large and enthusiastic listening audience."[29] Management said, "KSJO encountered great difficulty in judging Stone's program overall when employing the FCC indecency formulation, because the station was receiving strong local listener support for Stone (reflected in mail, telephone calls, and audience ratings)," at a time when Stone's show ranked number one among the 18 to 49-year-old men in the market.

Ultimately, KSJO management—which later sold the station—quickly moved away from shock radio:

> KSJO's experience with Stone led it to conclude several months in advance of the FCC inquiry that this type of humor is much too difficult to police and too offensive to certain members of the listening audience to warrant carriage on KSJO. The station, therefore, elected to abandon this type of programming . . . and has no intention of returning to it. In this regard, KSJO held meetings with certain complainants in May and June of 1989, publicly apologized to the station listeners who were offended by Stone's program, and made clear the change in direction of its morning program.[30]

KSJO-FM, on June 30, 1989, issued a "joint statement" with a group calling itself Coalition for Integrity in the Media (CIM). The KSJO-FM general manager signed the statement, which read in part: "We encourage community groups, especially those representing racial, ethnic, cultural, and sexual minorities to share their message through KSJO community programs and public service announcements."[31]

A tape log shows a pattern of content that went beyond sexual issues. Stone, according to his critics, had used "fighting words" against homosexuals by urging the killing or maiming of AIDS patients; he had employed antigay discrimination and AIDS hysteria; he had allowed callers to tell racist jokes "about shooting 'Africans,' 'Puerto Ricans,' and 'Mexicans,' " and he had also attacked Jews and Asians.

A variety of groups in the San Jose market challenged Stone and his show. The California Teachers Association, in a letter dated November 21, 1988, wrote that the board of directors had voted "to censure KSJO-FM for continuing to broadcast comments of disk-jockey Perry Stone."[32] The group argued that "freedom of expression is subject to legal, though minimal, constraints, "and likened these broadcasts to shouting "Fire!" in a crowded theatre:

> Mr. Stone's inflammatory rhetoric—vulgar and violent expressions of bias toward members of various racial and ethnic groups, toward handicapped people, people of differing sexual orientations, and practitioners of different religious faiths—is not merely in execrable taste. Socially, morally, and—we believe—legally, it is equivalent to the kind of speech Justice Holmes denied the protection of the Constitution.

The Women's International League for Peace and Freedom on November 17, 1988, sent a letter to KSJO-FM, and a copy to the FCC, concluding: "It is difficult to imagine a show more distasteful, repulsive, and dangerous to harmony and peace in our community, than your Perry Stone morning program."[33]

A number of individuals also placed comments in the public record. Said one San Jose resident, "When we think of some of the great humorists of America's past—Will Rogers, Charlie Chaplin, Mark Twain—where do we find this need to denigrate, condescend, belittle, and ridicule others for the sake of making a joke?"[34] The history professor asked station management to "tone down the tenor of remarks" made by Stone: "The hints of racism are more than hints; especially in the Bay Area are people quick to spot the difference between genuine, offbeat humor and a kind of radio show that is divisive, inflammatory, and devoid of constructive and positive social commentary."

A San Jose physician complained about a song Stone aired: "The words of the song included this quote, 'My wife ran off with a Nigger'. For some reason the deejay deleted the word Nigger as if by this omission the song would be more acceptable."[35] He added, "Any radio station given the privilege of using the public air-waves has a public responsibility to exercise that privilege in the public interest."

Another listener was outraged by a story about a Vietnamese restaurant in which a customer kills a delivery man: "By accident my radio was set on your station this morning and I heard 5 minutes of hate."[36]

The management of KSJO-FM was also facing coverage of the criticisms published in the *San Jose Mercury News* on September 8, 1988. "Eight months after Stone wise-cracked his way into San Jose, he has critics vowing an advertising boycott if he stays on, fanatic fans vowing a boycott if he's kicked off, and KSJO management worrying all the way to the bank." The newspaper quoted Stone:

> "I don't consider myself a prejudiced man by any stretch of the imagination," said Stone, 29, who projects a soft-spoken, respectful tone off the air. "I'm not using the air waves and saying I want you to go out and shoot 10 blacks today."
> Stone said he's "an equal opportunity offender: I make fun of myself. I make fun of the next person." Recently, Stone, an Italian-American, remade the pop tune "Pop Muzak" into a satire of Italian singers, "Wop Music."[37]

The newspaper also published a series of Letters to the Editor protesting that KSJO-FM was a racist rock station: "KSJO represents the worst side of the business world." Another listener wrote that the "rude racist remarks and sarcasm brought back some painful memories" of growing up in Ohio. However, one supporter of Perry Stone concluded that "if we can't laugh at ourselves, then we've got some serious troubles!" She added, "Perry Stone isn't forcing anyone to listen to his show." The FCC reports, however, that it never received any letters of support for the show.[38] The reluctance of management to act in the fall of 1988 led to a full-scale confrontation in the winter of 1989.

The community furor that had erupted in January, February, and March 1989 shows how an organized community effort can affect change in policy of a local station dependent upon the community for its economic livelihood.

Conclusions and Implications

The cases of WLUP-AM and KSJO-FM help illuminate the role radio audience and community members can play in the regulation process. Some generalizations are possible:

1. In both cases, some audience members were participants in the offending broadcasts, and in both cases station ratings seemed

to support the idea that there was interest in the type of programming being presented.

2.   In both cases, local newspaper pieces contributed to the public debate about indecency. However, these concerns were connected to broader concerns about broadcast racism.

3.   In the case of WLUP-AM, the complaints came from two individuals who had documented their evidence with audio tapes. In the case of KSJO-FM, the criticisms came from organized sectors of the society.

4.   The KSJO-FM broadcasts received little audience support, and no formal support (in terms of letters written to the FCC); but WLUP-AM marshaled a slew of letters from listeners that became part of the official record in the case.

5.   Management at WLUP-AM stood behind its broadcasts, but management at KSJO-FM ultimately embraced its critics and distanced itself from the offending broadcasts and broadcaster.

6.   Despite the differences in offending broadcasts, community and audience responses, and station reactions to the complaints, in both cases the regulatory results were similar: the FCC levied relatively small fines and did not threaten the licensees with loss of their profitable public franchises.

At the broader policy level, it seems clear that the regulatory process that occurs in Washington, D.C., appears to be rather distant from the public debate that occurs in a community where broadcast indecency is at issue. It seems awkward that the FCC is willing to act on complaint letters and tapes, but it is not willing to get involved at the local level. This historical policy decision may not serve the best interests of stations or the public they are licensed to serve.

The attempt by regulators to assess whether a complaint is actionable or not by applying a national standard against a community reaction has its limitations. Surely, there are many in a community who—even when the complaints are publicized—remain unaware. Local public hearings on complaints against stations, as well as on license renewal issues, might provide the backdrop for assessment of whether or not a station or program is serving the needs of the local audience.

On the other hand, the lack of local process functions quite well for local stations by keeping them out of the spotlight of local political controversy. As a First Amendment issue, the process probably keeps them free from some attempts to use the regulatory system as a means to stifle certain forms of free expression—whether they be sexual, political, or challenging to established order in other ways. The rather removed process makes good business sense in a regulatory system most concerned about the business aspects of broadcasting.

Regardless, the current process does not seem to satisfy offending broadcasters, their listeners who do not want expression chilled, their listeners who want such expression silenced, and licenses placed in more "responsible hands," or the larger community that may be offended by programs that promote unpopular attitudes.

Despite all of the controversy over broadcast indecency, the offending stations are few in number: most tote the inoffensive line, preferring to gravitate to the safe center of acceptable expression.

Manager's Summary

The cases of WLUP and KSJO reveal that broadcast managers wanting to sponsor shock jock's "raunch" radio should be prepared to defend their air personalities or suffer the consequences. Broadcast managers who see borderline broadcasts as financially viable must monitor the broadcasts and be involved in responding to community complaints.

Sample Letters of Complaint and Support

PALO ALTO COUNCIL OF PARENT-TEACHER ASSOCIATIONS · PALO ALTO, CALIFORNIA

RECEIVED BY

MAR 0 6 1989

FCC MAIL BRANCH   February 15, 1989

Federal Communications Commission
Mass Media Bureau
Enforcement Division
Complaints and Investigations Branch
1919 M Street, N.W.
Washington, D.C. 20554

Re: "The Perry Stone Show", KSJO-FM Radio, San Jose, California

As representatives of the Palo Alto (California) Parent-Teacher
Associations, we wish to bring to your attention our great concern about
the early morning broadcasts of the "Perry Stone Show" on radio station
KSJO-FM in San Jose, California. We have been moved to approach the FCC
on this issue due to our belief that the content of these shows is harmful to
the well-being of children.

Our special concern is for the children within the listening area of
KSJO who are uniquely impressionable and who have access to these
broadcasts due to the time of their airing (between 6 and 10 a.m.).
Children are a special class within our society, needing special protection
from potentially harmful circumstances where they cannot protect
themselves. Children have no built-in judgment and need to be protected
from exposure to language and subject matter which are morally
degenerative. It is our considered opinion that at a time in their lives
when children are most open to suggestion and vulnerable to negative
influences, these radio broadcasts constitute a clear and present danger to
their moral well-being and development.

While we respect the fundamental importance of freedom of speech
in a free society and its protection under the First Amendment, we believe
that interest is outweighed in this case by the overwhelmingly vulgar and
sexually provocative material broadcast during these programs. It is our
belief that the content of these radio broadcasts comes within the

FIGURE 6.1

prohibition against the utterance of obscene, indecent or profane language by means of radio communication contained in criminal statute 18 USCS §1464 and the interpretation of that statute with respect to children in F C C v. Pacifica Foundation, 438 U.S. 726 (1978).

We therefore respectfully request that you immediately investigate this matter and take all appropriate legal action to have these radio broadcasts moved to a time when children do not have access to them, or to have the offensive material discontinued.

We hope you will keep us fully informed of your actions in this regard.

Sincerely,

The Palo Alto Council of Parent-Teacher Associations
By

Charmaine L. Moyer

Charmaine Moyer, President

cc: Hon. Alan Cranston, U.S. Senate
Hon. Pete Wilson, U.S. Senate
Hon. Thomas Campbell, U.S. House of Representatives
Hon. Tom Lantos, U.S. House of Representatives
Hon. Norman Mineta, U.S. House of Representative
Board of Supervisors, Santa Clara County
Trustees of the Santa Clara County Unified School District
Sixth District, California Congress of Parents, Teachers and Students
Trustees of the Palo Alto Unified School District
Coalition for Integrity in the Media

FIGURE 6.1   *Continued*

November 21, 1988

Ken Anthony
KSJO-FM
1420 Koll Circle
San Jose, CA   95122

Dear Mr. Anthony,

At its meeting last weekend, the Board of Directors
of the California Teachers Association voted formally
to censure KSJO-FM for continuing to broadcast
comments of disk-jockey Perry Stone.

The Board also voted to seek the right to testify in
KSJO-FM's license-renewal hearing if Mr. Stone is not
removed from the station at once.

As the organizational representative of over 230,000
public school teachers, community college
instructors, and state university professors, CTA
believes strongly in the constitutional guarantee of
freedom of expression.

As responsible adults - responsible above all to our
students, the next generation of Americans - CTA
members also recognize that freedom of expression is
subject to legal, though minimal, constraints.

Justice Oliver Wendell Holmes noted that the right to
free speech does <u>not</u> include the right to shout
"Fire!" in a crowded theater.

Mr. Stone's inflammatory rhetoric - vulgar and
violent expressions of bias toward members of various
racial and ethnic groups, toward handicapped people,
people of differing sexual orientations, and
practitioners of different religious faiths - is not
merely in execrable taste. Socially, morally, and -
we believe - legally, it is equivalent to the kind of
speech Justice Holmes denied the protection of the
Constitution.

In a pioneering study titled <u>The Nature of Prejudice</u>,
sociologist Gordon Allport demonstrated a natural and
all but inevitable progression from slur to violence
- from "gentle" expressions of bias to bruising
insults of the Perry Stone character and beyond.

FIGURE 6.2

KSJO-FM
November 21, 1988
Page 2

If we have learned anything at all from the horrors
of the 20th century, we cannot - and must not - have
missed its central lesson: Intolerance of others,
because of their race, ethnic origins, physical
qualities, or religious beliefs, is a cancer on the
survival of the human race.

Mr. Stone's behavior is spreading the cancer. It
should be your responsibility - as I believe it is
that of educators and of all decent human beings - to
arrest that cancer.

Sincerely,

Ed Foglia
President

cc: D.A. "Del" Weber
    Ron McPeck
    Ralph J. Flynn
    Board of Directors
    Management Staff

FIGURE 6.2 *Continued*

REPLY TO:

☐   DISTRICT OFFICE
     100 PASEO DE SAN ANTONIO
     SAN JOSE CALIFORNIA 95113
     (408) 288-7515

☐   CAPITOL OFFICE
     STATE CAPITOL
     P.O. BOX 942849
     SACRAMENTO, CA 94249-0001
     TEL AREA CODE 916
     445-4253

COMMITTEES

EDUCATION
SUBCOMMITTEE ON
  POSTSECONDARY EDUCATION
ECONOMIC DEVELOPMENT &
  NEW TECHNOLOGIES

Assembly
California Legislature

JOHN VASCONCELLOS
ASSEMBLYMAN, TWENTY-THIRD DISTRICT

CHAIRMAN
COMMITTEE ON WAYS AND MEANS

RECEIVED BY

MAR 0 6 1989

FCC MAIL BRANCH

February 9, 1989

Coalition for Integrity in Media
152 S. 16th Street
San Jose, CA 95112

Friends -

I write in support of your FCC complaint protesting racist,
sexist and homophobic statements made on KSJO's "Perry Stone
Show". Please add my support in your supplementary mailing
to the FCC.

I believe that violence against gay men and lesbians,
against women, and against ethnic minorities is encouraged
by the unapologetic broadcast of such hostile stereotypes.
We must not allow this one irresponsible broadcast to impede
progress made by and in our community.

I wish you well.

FIGURE 6.3

Women's International League for Peace and Freedom
United States Section ● 1213 Race Street ● Philadelphia, PA 19107-1691 ● (215) 563-7110
SAN JOSE BRANCH, 3349 Weepingcreek Way, San Jose, CA 95121 (408) 274-3941

November 17, 1988

David Baronfeld
Manager, Radio Station KSJO
1420 Koll Circle
San Jose, Ca. 95112

Dear Mr. Baronfeld:

It is difficult to imagine a show more distasteful, repulsive and dangerous to harmony and peace in our community, than your Perry Stone morning program. The San Jose Branch of the Women's International League for Peace and Freedom urges you most vigorously to remove this insult to national and international standards of decency and respect from the air.

We feel that the media in our society have a responsibility to the community. In turn, community members must assume responsibility for monitoring and directing media's role—to insure democratic influence and expression in development of our cultural values. Values in our society are not accidentally acquired. They develop over time through customs, traditions, social relations, educational and historical development of many diverse elements. They are subject to every-day testing as guides for the survival of the society within its given economic framework. They are carefully monitored and nurtured as a measure of society's progress and health. Our media reflect, direct, influence these values. Your show violates the most positive values and achievements of our community toward racial harmony, toward healthy growth of our youth, toward tolerance, respect and understanding of society's diverse elements. Your show would tend to destroy the hard-fought-for goals of our educational and religious communities. It most clearly violates Federal Communications Commission standards of decency—defined as language describing sexual or excretory activities that is "patently offensive as measured by contemporary community standards" (Richard Bozzelli, FCC attorney). We reject the "values" promoted as a model by your "shock jock."

We list the types of features we find offensive:

● Misguided and obnoxious "humor" at the expense of different racial and ethnic groups in our community. This includes sarcasm, insult, ridicule, vulgarity, gross negative stereotyping and degradation of individuals, groups, accepted values and cultural practices. Women, religious groups, and gays are also victims of this sick "humor." There is a complete lack of respect and sensitivity for the rights and dignity of the various elements in our society, and for acceptable, honorable, and healthy moral standards.

● Scatological and sexual references and incitement that exploit and stimulate in an unhealthy direction awakening emotions and curiosity of impressionable listeners, particularly confused and immature youth who seek to emulate the "worldly adult." On this show, degradation of the human body and sex is their introduction to the adult world.

FIGURE 6.4

- Sensationalizing and romanticizing drug and criminal and amoral subcultures and exploiting revolting, vulgar, immoral, anti-social tendencies in individuals—personality components that, without healthy guidance, often lead to juvenile delinquency, and sex, drug, and inhumane crimes. There is incitement to violence and animal cruelty. Evil and violence are treated as "fun." Listener stories about putting urine into cologne bottles, and about group rape of a woman who had undergone a sex change operation, with Stone's encouragement of graphic and lurid details, are direct results of this kind of incitement.

- Violation of individual privacy: forcing naming of names and addresses.

- Encouraging and exposing small children as part of the participating audience.

Some blatant examples:

- Constant sickening, outrageous racial slurs against blacks, Asians, Mexican-Americans, gays, women, homeless, people with disabilities and illness, etc.

- God is a practical joke, etc.

- The "Sophie" character.

- Ridicule of Nancy's mastectomy, epileptics, blind people, etc.

- Ridicule of Oprah Winfrey.

- Breasts look like pimples. "Jokes" about uncircumcised penises.

- A caller tells of shooting <u>cans</u> —<u>Mexicans</u> and Puerto-<u>Ricans</u>.

- A beer enema. "Scummy" stories.

- Play kitty ball, using the cat as the ball.

- Suggestion to push people off cliff to collect their money.

- Suggestion of human sex with a cow by a caller.

- A free-for-all over football, ending with US vs. Canada hostility, with screaming verbal abuse, insult, and name-calling regarding merits and demerits of the respective countries.

San Jose has no place for this kind of immature madness, certainly not on public airways. Let us continue to build friendship and harmony among our many diverse groups, with mutual respect and consideration for all. Remove this foul obstruction to healthy growth and progress of our community!

Sincerely,

Dorothy Aspinwall
Dorothy Aspinwall
Chairperson

Copies to Federal Communications Commission, Santa Clara Human Relations Commission, Narragansett Broadcasting, San Jose Mercury News, Metro

FIGURE 6.4 *Continued*

EXHIBIT No. 2

J O I N T   S T A T E M E N T

KSJO FM-AND-COALITION FOR INTEGRITY IN THE MEDIA (CIM)

Radio Station KSJO FM and the Coalition for Integrity in the Media (CIM) are issuing this joint statement in the interest of promoting a community which values and respects all people and all cultures. We regret that any listener was offended by the now cancelled 'Perry Stone Show' and apologize to those persons and groups who were offended. In view of subsequent discussions, with many representative organizations whose opinions we value, KSJO FM has determined that the best interest of the community would be served by <u>not</u> reinstituting such programs in the future.

We encourage community groups, especially those representing racial, ethnic, cultural and sexual minorities to share their message through KSJO community programs and public service announcements.

The airwaves are a vital resource in our community; through our cooperative efforts, we will work to assure that they are used to build community awareness, community acceptance and community betterment.

_____   _____ 6-30-89
DAVID J. BARONFELD   MARJORIE BOEHM
General Manager   Chairperson
KSJO FM   CIM

JointSta

FIGURE 6.5

DALTON-FORD
Associates, Inc.

Mr. Larry Wert
General Manager
WLUP
875 N. Michigan Ave.
Chicago, IL 60611

September 8, 1989

Dear Mr. Wert:

The recent publicity surrounding the Steve Dahl and Garry Meier show have
prompted me to write this letter. A first for me to any radio or television
station. But I felt the circumstances demanded a response of my time.

I have been listening to the Steve and Garry Show for over ten years. I
personally feel that their show is one of the finest outlets and entertainment
available on the radio today. During the past ten years I seemingly spend
more time in my car than in my office and Steve and Garry are constant com-
panions, as well as other WLUP personalities. There are times when Steve
and Garry may push the edge of the creative envelope a little hard, but at
no point in the ten years I have listened have I ever heard anything over the
airwaves that I consider in violation of public standards. They are truly
a creative island in a sea of mundane stations.

Even though Steve and Garry are known primarily for their humor and irreverence,
there have been many times over the years that I feel they have offered
genuine service to our community by addressing topics that cannot find a
forum, or more importantly addressing topics that should be made aware to the
public. One of the first examples that leaps to mind is Steve Dahl's parody
song of the Rev. Ernest Angeley. This one song and subsequent discussions
brought into focus several years ago, the abuse of power and influence that
television ministers weild. As a direct result of the attention Mr.'s Dahl
and Meier gave this issue, the Reverend Angeley's practices were severely
curtailed. Including investigation by several Federal and Local authorities
that prompted Rev. Angeleys removal from several television stations. It does
not take a genius to figure out that Steve and Garry were ahead of their
time on this issue, all we have to do is turn our attention to the Jim Bakker
scandal to see the parallel's.

And more recently, Steve has been raising concerns over the Commonwealth
Edison promotion of their "E Team". The service that Com Ed provides to
school aged children that if they are in trouble they should go to a policeman
or a Commonwealth Edison lineman or even a Com Ed employee in one of their
tan cars. Mr. Dahl brought up the point that as a parent he did not want
teachers or other groups, ie. Com Ed, directing his children to strangers
in automobiles. And while the issue is more complicated than I have indicated
this is a program of which I was not even aware of, and because of the air
time devoted to this topic I learned about the program and even went so far
as to contact Com Ed regarding this issue. As a parent I too do not want
my children to be told to approach strangers, especially those in cars. Again,

852 Merchandise Mart Chicago, IL 60654 312.329.9394 312.467.1498

FIGURE 6.6

DALTON-FORD
Associates, Inc.

there is no other radio, or for that matter, television station that has
addressed this topic. I think Com Ed's position is admirable and is attempting to provide a service to the community but I still have some concerns.
I am so very happy that Steve and Garry brought this to my attention:

I could continue with the examples but I'm sure you get my point. Yes
there are times when some people in your listening audience may find the
Steve and Garry show unsuitable for their tastes. That is their perogative.
Certainly we have all learned that no one is going to appeal to every person
in our society. Whether you are speaking of an on-air personality or the
President of the United States.

One of the greatest elements of our society is our freedom of choice.   An
element that I think as Americans we sometimes tend to take for granted. If
any listeners contact your station regarding the content of Steve and Garry's
show, or any other show for that matter regarding its content, that is their
right as citizen's. However it is at their option that they tune into the
Loop. If they find what they are hearing is not for them, they have the
option of simply turning the dial to another station or turning the radio
off. I simply cannot understand the attitude that every one must conform
to every person's idea of fairness. If that is our attitude we might as
well pull off all entertainers and broadcast Muzak over the airwaves, but
I know some will object even to that.

My point Mr. Wert is that we have been seeing instances in our society that
one or two complaints from viewers or listeners can change the face of
American broadcasting. This is not operating in the public's best interest
when a disproportionate minority can dictate to the majority what it can or
cannot listen to.

As you can tell I feelvery strongly on the issue of free speech and felt that
I should take the time to compose this letter. If you would like to use this
communication for the FCC please feel free.

Thank you for your time and consideration in reviewing my letter. Please
do all that can be done to allow free speech, and Steve and Garry, to continue.
Thank you once again.

Very Truly Yours,

Mark G. Dalton

852 Merchandise Mart Chicago, IL 60654 312.329.9394 312.467.1498

FIGURE 6.6   *Continued*

18111 Wildwood
Lansing, IL 60438
31 August, 1989

Larry Wert
xWLUP AM 1000
875 N. Michigan
Chicago, IL 60611

Dear Mr. Wert:

I am writing to voice my support for Steve Dahl and Garry Meier. I have
been listening to Steve and Garry, following them around their various radio
homes, for over ten years. Not a day has gone by that Steve has not said
something that caused me to laugh to the point of hysterics. And more often
than laughter, he has inspired me to remark, "That's exactly what I have
been thinking, but was not able to put into words."

I resent this unwarranted attack on Steve and Garry for two reasons;
personal and constitutional. Personally, I take umbrage at the fact that
somebody who doesn't even know me is trying to dictate to what I may
listen. Also, I am outraged that my friends (and that is what I consider them
to be) are being persecuted on the whim of a tiny handful of non-fans. I
understand that there exists a small segment of society that does not like
their show (nobody that I know of), but that is understandable. I do not
like asparagus; so I do not eat it. I do not try to prevent others from eating
it. I just don't order it for myself. My friends and I, all adults and
professionals, have spent countless hours reminiscing about our favorite
Steve and Garry bits (my favorite is the Kup/Angeline Caruso fishing trip,
but that's not germane). More than three hundred thousand (300,000)
people love their show; do we define the arbitrary and capricious term
"community standards", or do the three people who wrote negative letters
define it?

The term "shock radio" has been bandied about lately. To me, that implies
gratuitous profanity and graphic descriptions. I have never heard Steve or
Garry swear out of context or for shock value. Steve is a master at relaying
his thoughts while making people re-think and re-evaluate their ideas.
Whether or not you agree with him, you will always be sure of your
reasons.

FIGURE 6.7

As for the constitutional arguments, something called the First Amendment comes to mind. Prior to the Reagan/Bush administration, I remember having the right to exercise free speech. An integral part of free speech is the freedom to listen. The freedom to <u>decide for ourselves,</u> based on our personal likes and dislikes, is part of being an American. Contributing to this right to exercise choice is the tuning knob that is available on many radios. By rotating this knob, one is able to choose among the variety of fare offered over the airwaves. If proof was available that Steve and Garry were Crazy-Gluing people's tuning knobs on AM 1000 and tying them to their chairs (like Gen. Dozier), I would fully support their being disciplined. But, lacking any such evidence, I can find no justification for separating them from their 300,000 loyal fans. If someone decides that Steve and Garry are not to their liking, they should exercise their constitutional right to <u>not</u> listen, just like I do whenever I happen to dial in Wally Phillips.

In closing, I fully support Steve Dahl and Garry Meier and their show. I can only hope that calmer heads will prevail and people will remember that this is America. At least it was when I woke up this morning.

Sincerely,

Lenny Munari

FIGURE 6.7   *Continued*

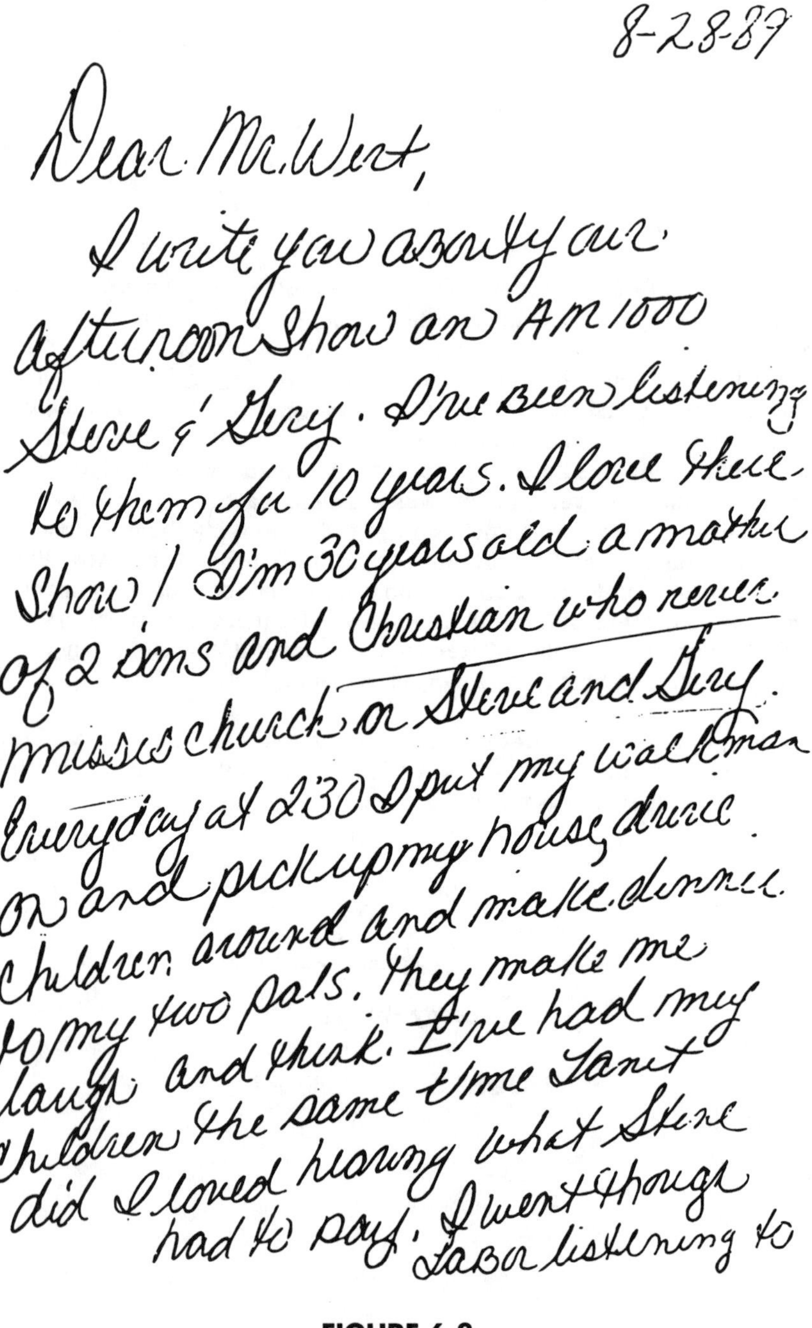

8-28-87

Dear Mr. West,

I write you about our afternoon Show an AM 1000 Steve & Gary. I've been listening to them for 10 years. I love there Show! I'm 30 years old a mother of 2 sons and Christian who never misses church on Steve and Gary. Everyday at 2:30 I put my walkman on and pick up my house, drive children around and make dinner to my two pals. They make me laugh and think. I've had my children the same time Janet did I loved hearing what Steve had to say. I went though Labor listening to

FIGURE 6.8

September 5, 1989

Federal Communications Commission
Mass Media Bureau
Enforcement Division

Attention: Complaints and Investigations Branch

    I was greatly disturbed to read that WLUP-AM of Chicago
was facing possible disciplinary action from the Federal Communications
Commission for broadcasting allegedly offensive descriptions
and depictions during the "Steve Dahl and Garry Meier" programs
of March 30, 1989 and August 19, 1987. With its recent issuances
of warnings to radio stations that practice so-called "shock
radio" , the Commission has demonstrated that it wishes to curb
the airing of gratuitous and mindless obscenities. As a regular
listener to the "Steve and Garry" program, I full believe
such charges against this radio show are completely unwarranted.

    Certainly there are references and remarks made on the show
that will strike some of the more sensitive listeners as excessive
or offensive. Of course, there has always been a portion of
the public that strongly opposes the usage of language they find
personally offensive in every form of mass comminucatioh. James
Joyce's Ulysses, the novels of D.H. Lawrence, and the works of the
great satirists from Jonathan Swift to Mark Twain to Thomas
Pynchon have all been the subject of debate at to whether their
use of scatological descriptions and vulgar language invalidates
their artistry, and the overwhelming concord of opinions by
judges, critics, and the general public has been that the author's
perception of truth should not be censured because of the potential
for offense. Comparing the writings by acclaimed artists to
the conversations of a couple of radio personalities may seem
absurd to some, but the work of Dahl and Meier is just as important
to the advancement of the medium as were the forementioned writers
significant in progressing their form of expression, often in
language far more impure than that used by these broadcasters.

    Before any action is taken by the Commission, I urge you
to consider that Steve Dahl and Garry Meier are among the very
few broadcasters in Chicago who take their oppurtunity to communicate
with thousands of people seriously. Rather than spout typical
dee-jay idiocies or reduce the radio into an information booth
for the time, temperature and location of clogged expressways,
Dahl and Meier have chosen to talk about what most people

FIGURE 6.9

Complaints and Investigations Branch Page 2 September 5, 1989

discuss, in the manner most people discuss things. In comparison
to most other programs, this is a very courageous stance. If
there is black humor on the show, it is quite a normal reaction
to what goes on in our not always pretty, not always nice world.

I assume there has been some concern that filth is being
broadcast and that children may hear it. An examination of the
ratings demonstrates that the vast majority of listeners to the
"Steve and Garry" program are in their mid-twenties to mid-thirties.
To force the show to suppress its biting satire would be an insult
to the regular listeners who are intelligent and mature enough
to enjoy the content of the program, and are not sitting by the
radio hoping Steve and Garry will start talking dirty to them.
Although my knowledge of the regulations is limited, it does not
seem to me that Dahl and Meier have been offensive to the average
listener, and to penalize them would be as grossly unfair as
banning books in the public library by Swift, Twain, or Pynchon
simply because a child may stumble upon a "smutty" passage.

The freedom of expression is as sacred to me as I'm sure
it is to you. To crackdown on responsible broadcasters such as
Steve Dahl and Garry Meier will strip radio of its ability
to address its listeners on the level of sophistication they
want and deserve. As playwright David Mamet put it, "Freedom
of speech (is) tolerated only until its exercise is found offensive,
at which point...freedom (is) haughtily revoked." Do not let radio
become a monument to banality.

Respectfully,

[signature]

John Wendler
632 Harrison
Oak Park, Il.
60304

FIGURE 6.9 *Continued*

Notes

1. *Pacifica Foundation*, 56 FCC 2d 94 (1975).
2. 932 F.2d 1507.
3. See T. Barton Carter, Marc A. Franklin, and Jay B. Wright, *The First Amendment and the Fifth Estate: Regulation of Electronic Mass Media*, 2nd ed. Westbury, N.Y.: The Foundation Press, 1989, pp. 276–345. The *NBC* case, as a matter of broadcast history, has squarely placed responsibility in the hands of licensees. See Christopher H. Sterling and John M. Kittross, *Stay Tuned: A Concise History of American Broadcasting*, 2nd ed. Belmont, Calif.: Wadsworth, 1990, p. 237, citing *National Broadcasting v. United States*, 319 U.S. 190 (1943). The FCC has the power to regulate beyond being a mere "traffic cop" of the airwaves. "It puts upon the Commission the burden of determining the composition of the traffic."

4. Sterling and Kittross, p. 88: "In the earliest years, self-regulation meant little other than 'silent nights.' Time-sharing was mostly voluntary, but . . . there was always threat of government action. In technical matters, broadcasting clearly needed a governmental traffic cop; in programming, broadcasting managed alone, typically exercising its freedom by censoring many dissenting points of view and presenting a conservative, business-oriented middle-class viewpoint." The industry, in more recent years, has relied upon the National Association of Broadcasters, including its statements about social responsibility, to deflect criticism about offending broadcasters. As an industry, there is an attempt to portray offenders as not representing the larger group (p. 364).

5. Jeremy Harris Lipschultz, "A Content Analysis of Broadcast Indecency Non-actionable Material," poster session paper, Boadcast Education Association, Radio '91, San Francisco, September 1991.

6. See *Yale Broadcasting v. FCC*, 478 F.2d 594 (D.C. Cir. 1973); *In Re WUHY-FM Eastern Educational Radio*, 24 FCC 2d 408 (1970); *In Re Apparent Liability, WGLD-FM*, 41 F.C.C. 2d 919 (1973); *Sonderling*, 41 F.C.C. 2d 777 (1973); *Illinois Citizens Committee for Broadcasting v. FCC*, 515 F. 2d 397 (D.C. Cir. 1974); and *Infinity Broadcasting*, 2 FCC Rcd. 2705 (1987). Cf. *Miller v. California*, 413 U.S. 15 (1973). The case establishes a contemporary community standards test for review of patently offensive sexual material that lacks social value in determining unprotected "obscenity."

7. The language originated with the Federal Radio Commission in 1928: "In a sense a broadcasting station may be regarded as a sort of mouthpiece on the air for the community it serves. . . ." cited in Federal Communications Commission, "Public service responsibility of broadcast licensees: Some Aspects of 'Public Interest' in Program Service," in Bernard Berelson and Morris Janowitz, p. 233, 238–239. Among public issue problems noted was: "12. Is a denial of free speech involved when a commentator is discharged or his program discontinued because something which he has said has offended (a) the advertiser, (b) the station, (c) a minority of his listeners, or (d) a majority of his listeners?" No "categorical answers" were offered. "Rather than enunciating general policies, the Commission reaches decisions on such matters in the crucible of particular cases."

8. William B. Ray, FCC, *The Ups and Downs of Radio-TV Regulation*, Ames: Iowa State University Press, 1990, p. 71.

9. 852 F.2d 1341. In a Santa Barbara case on the fitness of a 10 P.M. broadcast, an estimated 1,200 children between the ages twelve and seventeen were in the audience during the 7 P.M. to midnight Saturday period. Such ratings, however, fail to suggest a magic hour when children are no longer present.

10. Letter to James de Castro, president, Evergreen Media Corporation of Chicago, licensee, WLUP-AM, 8310-MD.

11. Ibid., p. 1. The station response to a 24 August letter of inquiry was dated 10 October 1989 to Edythe Wise, Chief, Complaints & Investigations Branch, Enforcement Division, Mass Media Bureau.

12. Ibid., p. 2.

13. Ibid., p. 2.

14. *Sable Communications of California, Inc. v. FCC*, 109 S.Ct. 2829 (1989). Protection of children was seen as "a compelling governmental interest" not requiring

proof of harm; Notice of Apparent Liability for a forfeiture, 8310-MD, 30 November 1989, p. 2. Section 1464 provides: "Whoever utters any obscene, indecent or profane language by means of radio communication shall be fined not more than $10,000 or imprisoned not more than two years, or both." Justice Stewart's dissent in *Pacifica* noted that the meaning of the term "indecent" was statutorily "limited to the sort of 'patently offensive representations or descriptions of that specific "hard core" sexual conduct given as examples in *Miller v. California*.' " The term "indecent" has often been confused with "obscene" in the broadcast context. Justice Stewart found that the legislative history shed no light on the meaning of "indecent" broadcasts: "Neither the committee reports nor the floor debates contain any discussion," 438 U.S. 726, fn. 5. Justice Stewart concluded that Congress intended no more than prohibition of obscene speech.

15. See Joe Flint, "Evergreen to fight indecency charge, since it has no avenue of appeal for FCC fine, it will refuse to pay; matter then gets handed over to Justice Department," *Broadcasting*, 13 January 1992, p. 91. The station could be charged with indecency in federal district court. Evergreen has concluded that "the commission's policies on broadcast indecency are unconstitutionally vague and unworkable."

16. FCC letter to Robin L. Shaffert, Latham & Watkins, Washington law firm for Evergreen Media (WLUP), in response to FOIA Control No. 89-155, dated 29 September 1989. "These documents . . . generally consist of eight, handwritten letters of complaint with seven accompanying tape recordings of WLUP broadcasts and one typewritten letter of complaint with an additional tape recording" (at p. 2).

17. Ibid., Appendix (1) (d). A one-page typewritten undated letter of complaint was received by the FCC 13 December 1988, and it included an undated tape recording. That letter read as follows:

> F.C.C.
> Mass Media Bureau
> Complaints
> Washington, D.C.
> 20554
>
> Dear F.C.C.,
>
> I want to start by saying I've never written to you before. I don't do this kind of thing often. Something must be done about Steve Dahl & Gary Muledeer. I've enclosed this tape of only 25 minutes of their show. It is worse than obscene, if there is such a thing. This type of thing goes on every day.
>
> I am a totally liberal person, but this must stop. It isn't played, or broadcasted, overnight. This junk is on at 5 p.m. If you can't do anything about what's on this tape, then how about this:
>
> 1. He said the day Dan Quale was in town someone should hit him with their car.

2. They promote a Child-molesting Boys Camp every day they're on the air. I've called the F.C.C. in Chicago, but nothing has been done. Please there must be something you can do. This is a disgrace to our Country. I'm not saying take them off the air, but at least put them in the middle of the night so our children can't hear. I am afraid of leaving my name and address because Steve Dahl has read complaints on the air. [Deleted]

Thank you,
[Deleted]

A similar letter in Appendix (1)(a), dated 21 August 1987, was sent to a regional F.C.C. director in Park Ridge, Illinois. That letter identified WLUP's discussion of masturbation, and use of words such as "penis" and "vagina": "I can't believe that you allow this to go on," the anonymous complainant wrote, "please consider this a complaint." There is no evidence on file that the regional office took any action against the station in 1987.

18. Ibid., Appendix (1)(b).
19. Ibid., citing Charles Chi Halevi, "Anti-Semitic jokes on radio, TV aren't one bit funny," Commentary, *Chicago Sun-Times*, 25 March 1988, p. 44.
20. The fifty-five-page station response drew from the exhibit letters as representing "average yet very diverse members of the Chicago community," 10 October 1989, p. 31.
21. Ibid., p. 18.
22. Letter to WLUP General Manager Larry Wert, dated 8 September 1989.
23. Letter to Larry Wert, dated 31 August 1989.
24. Another handwritten note.
25. One may argue organized group comment for or against a station might weigh more heavily as a political force as the FCC determines how it should handle a particular case. In its first notice to KSJO-FM, the Mass Media Bureau wrote: "This is in reference to a letter dated January 12, 1989, from a group of individuals and organizations from San Jose, CA, complaining of the broadcast of allegedly indecent language . . . ," 24 August 1989.
26. Ibid., pp. 3–6.
27. In 1991, Narragansett Broadcasting Co. of California Inc. sold KSJO-FM and KSJX-AM to Baycom Partners Ltd. for a reported $5.4 million. See *Broadcasting*, "Special Report: Station Trading 1991," 10 February 1992, p. 26.
28. Peter D. O'Connell, Reed Smith Shaw & McClay, including Pierson, Ball & Dowd, Washington, D.C., letter dated 25 September 1989, p. 2.
29. Ibid., p. 3, and fn. 3
30. Ibid., pp. 4–5, and fn. 5.
31. Exhibit No. 2, Joint Statement, signed by David J. Baronfeld, General Manager, KSJO-FM, and Marjorie Boehm, Chairperson, CIM, dated 30 June 1989.
32. Ed Foglia, president, letter to Ken Anthony, dated 21 November 1988.
33. Dorothy Aspinwall, chairperson, letter to David Baronfeld with copies to the FCC, the Santa Clara Human Relations Commission, Narragansett Broadcasting, and the *San Jose Mercury News*, dated 17 November 1988.
34. Dr. Roger E. Rosenberg, San Jose, letter to KSJO, dated 15 September 1988, pp. 2–3.

35. Richard Turner, M.D., letter to Phil Norton, dated 27 January 1988.
36. Wendy Lieber Poinsot, Palo Alto, letter to Kay Hickey, KSJO-FM, dated 19 July 1988.
37. *San Jose Mercury News*, 8 September 1988, p. 18A.
38. Roy J. Stewart, Chief, Mass Media Bureau, Federal Communications Commission, letter to author, dated 1 November 1991: "Initially, please be advised that a review of the record systems maintained by the Complaints and Investigations Branch of the Mass Media Bureau has not located any letters in support of the KSJO-FM, San Jose, California broadcast to which you refer." See William Steel, Redwood Estates, Letters to the Editor, "Protest San Jose's racist rock station," *San Jose Mercury News*, 17 October 1988; Rick Graves, San Jose, Letters to the Editor, "Stone defames the ethnic majority," *San Jose Mercury News*, 18 October 1988; and Alice Ralston, San Jose, Letters to the Editor, "Stone's irreverence lightens daily load," *San Jose Mercury News*, 18 October 1988.

Chapter 7

Branton v. FCC:
The Redefinition of Listener Standing

In 1993 the United States Court of Appeals, District of Columbia Circuit, issued an opinion in the case of *Branton v. FCC* that may significantly affect the ability of broadcast audience members to influence FCC regulation of stations.[1] The purpose of this chapter is to review the decision and its implications for broadcasters.

Branton Issues: The Legal Role of Audience Members

At issue was whether an "offended" listener has legal standing to challenge FCC inaction in an indecency complaint.[2] Peter Branton had complained about a 1989 National Public Radio *All Things Considered* broadcast of a profanity-filled tape from the John Gotti trial. NPR chose not to edit or cover language contained in the wiretapped telephone conversation between the alleged mob leader and another man. The FCC found the broadcast to be covered under "bona fide news." The court dismissed Branton's appeal.[3] A portion of that decision, which the United States Supreme Court let stand, is reproduced below:

Peter BRANTON, Petitioner v. FEDERAL COMMUNICATIONS COMMISSION and United States of America, Respondents

Radio-Television News Directors Association, et al., Intervenors

No. 91-1115

UNITED STATES COURT OF APPEALS FOR THE DISTRICT OF COLUMBIA CIRCUIT

September 9, 1992, Argued

June 1, 1993, Decided

Before BUCKLEY, WILLIAMS, and D.H. GINSBERG, Circuit Judges.

Opinion for the court filed by Circuit Judge D.H. GINSBERG.

I. Factual Background

In the early evening of February 28, 1989, NPR's news show "All Things Considered" ran a report on the trial of John Gotti, the alleged leader of an organized crime syndicate in New York. The report featured a tape recording of a wiretapped phone conversation between Gotti and an associate. In the 110-word passage that NPR excerpted from the tape recording for broadcast, Gotti used variations of "the f--- word" ten times. He used it to modify virtually every noun and in one instance even a verb ("I'll f---ing kill you"). NPR made no effort, such as substituting bleeps for any or all of these references, to render the passage less offensive to persons of ordinary sensibility.

Peter Branton, who heard the broadcast and was offended, filed a complaint with the Mass Media Bureau of the FCC. The Bureau concluded that the broadcast material in question was "not actionably indecent" and did not provide "the necessary legal basis for further Commission action" pursuant to *18 U.S.C. @ 1464*. Mr. Branton then wrote to the Commission asking how he could appeal the Bureau's decision. The Commission treated his letter as an Application for Review and, in a brief letter ruling (over one dissent), affirmed the Bureau's decision. The Commission explained that the Gotti tape was part of a "bona fide" news story; indeed, it had been introduced as evidence in the criminal trial that was the subject of that story. The Commission also noted its long-standing reluctance "to intervene in the editorial judgments of broadcast licensees on how best to present serious public affairs programming to their listeners." *Letter Ruling, 6 FCC Rcd. 610 (1991).*

Mr. Branton now petitions for judicial review of the agency's decision not to proceed against NPR.

II. Analysis

Article III of the *Constitution of the United States* limits the scope of the federal judicial power to the resolution of "cases" or "controversies." In order to implement that limitation, the Supreme Court has developed a doctrine of standing that, along with the other requirements for justiciability, assures that the federal judicial power is exercised only in "those disputes which confine federal courts to a role consistent with a system of separate powers and which are traditionally thought to be capable of resolution through the judicial process." . . .

In order to establish standing under *Article III*, a complainant must allege (1) a personal injury-in-fact that is (2) "fairly traceable" to the defendant's conduct and (3) redressable by the relief requested. . . .

The alleged injury must be "distinct and palpable," . . . not "conjectural" or "hypothetical," *Los Angeles v. Lyons*, 461 U.S. 95, 101–02, 75 L. Ed. 2d 675, 103 S. Ct. 1660 (1983). Application of these familiar principles leads us to conclude that the petitioner lacks standing to seek review of the FCC no-action letter at issue here.

A. Injury-in-fact: In order to challenge official conduct one must show that one "has sustained or is immediately in danger of sustaining some direct injury" in fact as a result of that conduct. *Golden v. Zwickler*, 394 U.S. 103, 109, 22 L. Ed. 2d 113, 89 S. Ct. 956 (1969). This component of the standing doctrine serves both "to assure that concrete adversariness which sharpens the presentation of issues," *Baker v. Carr*, 369 U.S. 186, 204, 7 L. Ed. 2d 663, 82 S. Ct. 691 (1962), and to prevent the federal courts from becoming "continuing monitors of the wisdom and soundness of Executive action. . . ." *Allen v. Wright*, 468 U.S. at 790 (quoting *Laird v. Tatum*, 408 U.S. 1, 15, 33 L. Ed. 2d 154, 92 S. Ct. 2318 (1972)).

The petitioner in this case alleges that he was injured because he was subjected to indecent language over the airwaves. While an offense to one's sensibilities may indeed constitute an injury, see *FCC v. Pacifica*, 438 U.S. 726, 748–49, 57 L. Ed. 2d 1073, 98 S. Ct. 3026 (1978), a discrete, past injury cannot establish the standing of a complainant, such as Branton, who seeks neither damages nor other relief for that harm, but instead requests the imposition of a sanction in the hope of influencing another's future behavior. "Past wrongs do not in themselves amount to that real and immediate threat of injury necessary to make a case or controversy." *Lyons*, 461 U.S. at 103. *See also O'Shea v. Littleton*, 414 U.S. 488, 495–96, 38 L. Ed. 2d 674, 94 S. Ct. 669 (1974). ("Past exposure to illegal conduct does not in itself show a present case or controversy regarding injunctive relief . . . if unaccompanied by any continuing, present adverse effects.")

If the petitioner suffers any continuing injury, we suppose it is in the nature of the increased probability that, should the NPR broadcast go unsanctioned, he will be exposed in the future to similar indecencies over the airwaves. Under established Supreme Court precedent, however, this marginal increase in the possibility of a future harm does not meet the "immediacy" requirement for *Article III* standing. For example, in *Los Angeles v. Lyons*, 461 U.S. at 95, the Court held that a person injured when a policeman subjected him to a chokehold did not have standing to seek an injunction prohibiting the police department from using that maneuver in the future. The Court reasoned that the plaintiff's single experience with a chokehold did not establish "a real and immediate threat that he would again be stopped for a traffic violation, or any other violation, by an officer or officers who would illegally choke him. . . ."

In the present case, the possibility that the petitioner will again "some day" be exposed to a broadcast indecency lacks the imminence required under *Lyons*, *Rizzo*, and *Defenders of Wildlife*. It is mere conjecture that a radio station will again broadcast, at a time when the petitioner is listening, indecencies that would be proscribed under *18 U.S.C. @ 1464* (as he would have us interpret that statute). While there is, of course, some chance that somewhere, at some time, the petitioner may again be exposed to a broadcast indecency as a result of the Commission's decision, that possibility seems to us far too remote and attenuated to establish a case or controversy under *Article III*.

Nothing in *Office of Communication of United Church of Christ v. FCC*, 123 U.S. App. D.C. 328, 359 F.2d 994, 1005–06 (D.C. Cir. 1966) *(UCC)*, is to the contrary. In that case, we held that "responsible and representative groups" in a broadcaster's listening area have standing to challenge the broadcaster's application for a renewal license. The appellants there alleged that a TV licensee had failed to "give a fair and balanced presentation of controversial issues, especially those concerning Negroes," in violation of the Fairness Doctrine and of the licensee's obligation to operate its station in the public interest. *Id. at* 1000 & 998–99. The court explained that "since the concept of standing is a practical and functional one designed to insure that only those with a genuine and legitimate interest can participate in a proceeding, we can see no reason to exclude those with such an obvious and acute concern as the listening audience." *Id. at* 1002.

UCC is not controlling in the present case for two reasons. First, the appellants in *UCC* alleged that the licensee in question was engaged in a continuing pattern of inappropriate and discriminatory broadcasting, which the FCC by renewing its license had in effect extended. In contrast, the appellant in the present case challenges an FCC determination regarding an isolated indecency broadcast at a single moment in the past. He does not allege a continuing course of misconduct, and there is simply too little reason to believe that the harm to him will ever recur. A listener who alleges that a broadcaster has repeatedly violated the indecency standard might be in a better position to argue that he is subject to a continuing harm or at least an increased likelihood of the harm recurring. But cf. *Lyons*, 461 U.S. at 105 (allegation that Los Angeles police "routinely apply chokeholds" fails to establish prior victim's standing).

Second, in the years since *UCC*, the Supreme Court has repeatedly emphasized the "immediacy" element of the injury-in-fact requirement. See *Defenders of Wildlife*, 112 S. Ct. at 2130 (1992); *Lyons*, 461 U.S. at 96 (1983); *Rizzo*, 432 U.S. at 362 (1976); *O'Shea*, 414 U.S. at 488 (1974). Accordingly, *UCC* must be understood as a creature of the context from which it arose, viz. a license renewal proceeding, which is inherently future oriented.

In sum, the petitioner fails to demonstrate that the FCC's decision not to take action against NPR causes him an injury that is sufficiently "immediate" to establish his standing to challenge that decision. The marginal increase in the probability that he will be exposed to indecent language in the future if NPR is not sanctioned is simply too slight to generate a case or controversy proper for resolution by an *Article III* court.

B. Causation/Redressability: Even if the harm to the petitioner here were sufficiently immediate to make out an *Article III* "injury," he would not be able to show that his injury "fairly can be traced to the challenged action" and would be "redressed by a favorable decision."

As an initial matter, it should be remembered that the petitioner in the present case is not himself in any way subject to the FCC decision he seeks to challenge. His probabilistic injury (such as it is) "results from the independent action of some third party not before the court," i.e., NPR or some other broadcaster that may in the future offend him with an indecent broadcast. *Simon*, 426 U.S. at 42.

In the present case, it is at least equally conjectural whether the FCC's proceeding against the alleged broadcast indecency of February 28, 1989, would cause any radio station(s) in the petitioner's area to broadcast any fewer indecent programs in the future. As in the cases discussed above, any favorable impact of the official action that the complainant seeks to compel depends utterly upon the actions of "third parties not before the court," *Simon*, 426 U.S. at 42, whose behavior is difficult to predict. For example, radio stations might well decide that the benefits of broadcasting indecent language of the sort petitioner here challenges outweigh the costs of making certain payments to the Government (here in the form of fines rather than of taxes). Predicting the reaction of "public" radio stations to a monetary fine is particularly difficult because such stations are non-profit entities. See Henry B. Hansmann, *Reforming Nonprofit Corporation Law*, 129 U. Pa. L. Rev. 497, 568–69 (1981) ("the patrons of a nonprofit are generally much less able to look out for themselves than are the shareholders in a business corporation"). In addition, as broadcast journalists, even for-profit broadcasters undoubtedly make programming decisions in part with an eye to non-monetary factors, such as their own conception of journalistic integrity. See *Radio-Television News Directors Association Code of Ethics* (stating members' responsibility, inter alia, "to gather and report information of importance and interest to the public accurately, honestly and impartially" and to "evaluate information solely on its merits as news").

As a result, the court can have no confidence that the FCC's failure to impose a sanction upon NPR will lead it or any other broadcaster to injure the petitioner in the future. Or to put the matter conversely, it is speculative whether our reversal of the agency's decision would serve at all to protect the petitioner from future exposure to broadcast indecency.

This holding (and the Supreme Court precedent upon which it is based) may at first seem inconsistent with the fundamental principle that increasing the price of an activity (i.e., broadcasting indecency) will decrease the quantity of that activity demanded in the market, or in the language of economics, that demand curves are downward sloping. See Paul A. Samuelson & William D. Nordhaus, *Economics* 60–61 (12th ed. 1985); Richard A. Posner, *Economic Analysis of Law* 5 (4th ed. 1992). We would hardly undertake to doubt this basic principle, however. See *Smith v. NTSB*, 981 F.2d 1326, 1328 (D.C. Cir. 1993). Rather, our concern is with the magnitude of its effect in this particular case (i.e., with the elasticity of demand for broadcast indecency). Without some reason to believe that the level of broadcast indecency is significantly affected by the possibility of incurring an FCC sanction, we lack a sufficient basis for the exercise of the federal judicial power. A court is rightly reluctant to enter a judgment which may have no real consequence, depending upon the putative cost-benefit analyses of third parties over whom it has no jurisdiction and about whom it has almost no information.

III. Conclusion
This dispute between the petitioner and the FCC falls outside the constitutional domain of the federal courts. The petitioner fails to establish a justiciable case or controversy because his asserted injury is too attenuated and improbable and because this injury neither resulted from the challenged Government decision nor would be remedied by a reversal of that decision.

Accordingly, the petition for review is Dismissed.

Narrow Legal Reading

The court read the law of legal standing narrowly by finding that "personal injury-in-fact" must be "fairly traceable" to "conduct" and "redressable" by the action sought.[4]

In the *Branton* case, Peter Branton claimed that "he was injured because he was subjected to indecent language over the airwaves."[5] Citing *Pacifica*, the *Branton* court said that while "offense to one's sensibilities" may be at issue in an indecency complaint, that is not enough. In the words of the court: "[A] discrete, past injury cannot establish the standing of a complainant, such as

Branton, who seeks neither damages nor other relief for that harm, but instead requests the imposition of a sanction in the hope of influencing another's future behavior."[6] In short, the court requires for legal standing a claim of actual injury:

> If the petitioner suffers any continuing injury, we suppose it is in the nature of the increased probability that, should the NPR broadcast go unsanctioned, he will be exposed in the future to similar indecencies over the airwaves. Under established Supreme Court precedent, however, this marginal increase in the possibility of a future harm does not meet the "immediacy" requirement for *Article III* standing.[7]

The subtle distinction is that in *Pacifica* the listener claimed he and a child had been injured by the 2 P.M. broadcast of the Carlin monologue. In effect, the decision ignores *Office of Communication of United Church of Christ v. FCC*, which had been the media case defining the issue of listener standing.[8] *OCC* granted "responsible" and "representative groups" of listeners legal standing at license renewal. The *Branton* court, however, ruled that *UCC* should not be applied because:

1. *UCC* involved "a continuing pattern of inappropriate and discriminatory broadcasting." Branton "challenges an FCC determination regarding an isolated indecency broadcast at a single moment in the past. He does not allege a continuing course of misconduct, and there is simply too little reason to believe that the harm to him will ever recur. A listener who alleges that a broadcaster has repeatedly violated the indecency standard might be in a better position to argue that he is subject to a continuing harm or at least an increased likelihood of the harm recurring."[9]
2. A series of recent Supreme Court cases stress "immediacy" as an element in determining injury. Because *UCC* was a license renewal case, the *Branton* court argues it was "inherently future oriented," and could not meet the new test. Peter Branton, likewise, made such a challenge.[10]

The *Branton* court rejects as "conjectural" the idea that an indecency fine or other action would affect other broadcasts in the future.[11] The court suggested that the behavior of other stations, not involved in a pending indecency complaint, "is difficult to predict."[12] For example, radio stations might well decide that the benefits of broadcasting indecent language of the sort petitioner here challenges outweigh the costs of making certain payments to the Government (here in the form of fines rather than of taxes). Predicting the reaction of "public" radio stations to a monetary fine is particularly difficult because such stations are non-profit entities.[13]

The *Branton* decision appears to make it very difficult for an audience member to show legal standing in a case outside of license renewal, where no immediate injury can be proven.

While the FCC routinely uses listener and viewer complaints—as well as station disclosure—as relevant information, there is no legal burden on the FCC or the courts to treat an audience member as an affected party in the decision. Audience members, under *Branton*, have no right to appeal a FCC decision to dismiss a complaint.

In these terms, regulation of broadcast indecency on the airwaves would seem to be a regulatory matter between the FCC, with its power to apply "public interest" standards, and offending stations.

FCC Decision in the Branton Case (1991)

Peter Branton
Report No. 44-9
FEDERAL COMMUNICATIONS COMMISSION
1991 FCC LEXIS 3141 RELEASE-NUMBER: FCC 91-27
(38103)
January 25, 1991 Released; January 24, 1991

Opinion:

Mr. Peter Branton
1007 Scenic Highway
Lookout Mountain, TN 37350

Dear Mr. Branton:

This letter concerns your November 17, 1989, *Petition for Reconsideration* of the Mass Media Bureau's October 26, 1989, dismissal of your complaint against National Public Radio. That request is being treated as an Application for Review pursuant to Section 1.115 of the Commission's Rules, given your expressed desire for eventual full Commission review. Your complaint alleged that a segment about John Gotti, a reputed organized crime figure, broadcast February 8, 1989, at 6:25 p.m. on the National Public Radio program "All Things Considered," was indecent. Specifically, the recording you supplied included, as part of a news segment about organized crime, a wiretap of a telephone conversation in which Gotti repeatedly used variations of the expletive "fuck" (transcript attached). In your Petition, you claim that it is difficult to understand how the language in the broadcast can be considered less than in violation of any standards of "obscene or indecent material." You also state the broadcast contained a preliminary admission that the language is "extremely profane" and note that an "obscene word" was repeated numerous times in the broadcast.

"Indecent programming" is defined as programming "that describes in terms patently offensive as measured by contemporary community standards for the broadcast medium, sexual or excretory activities or organs." *Infinity Broadcasting Corp.*, 3 FCC Rcd 930 (1987), aff'd in part and remanded in part sub nom. *Action for Children's Television v. FCC*, 852 F.2d 1332 (D.C. Cir. 1988).

When analyzing whether particular material meets this definition, the Commission looks first and foremost to the context in which the language was presented—a review that encompasses a host of variables, including, among other things, an assessment of whether the language was used in a "shocking" or "vulgar" fashion or was without merit. *Id.* at 932. No terms are per se indecent, and words or phrases that may be patently offensive in one context may not rise to the level of actionable indecency if used in other, less objectionable circumstances. *Id.* at 932 and n.28; *see also* Letter from James C. McKinney to William J. Byrnes (June 5, 1987), *aff'd Memorandum Opinion and Order*, FCC 87-215 (June 16, 1987).

We recognize that the repetitious use of coarse words is objectionable to many persons, and understand that you personally may have been offended by the use of expletives during the Gotti segment. Nonetheless, we do not find the use of such words in a legitimate news report to have been gratuitous, pandering, titillating or otherwise "patently offensive," as that term is used in our indecency definition. In reaching this determination, we note that the program segment, when considered in context, was an integral part of a bona fide news story concerning organized crime, and that the material prompting your complaint was evidence used in a widely reported trial. *See Infinity Broadcasting*, 3 FCC Rcd at n.31. These surrounding circumstances persuade us that the use of expletives during the Gotti segment does not meet our definition of broadcast indecency. In reaching this conclusion, we note that we traditionally have been reluctant to intervene in the editorial judgments of broadcast licensees on how best to present serious public affairs programming to their listeners. *See, e.g., Syracuse Peace Council*, 2 FCC Rcd 5043, 5051 (1987), *recon. denied*, 3 FCC Rcd 2035 (1988), *aff'd sub nom. Syracuse Peace Council v. FCC*, 867 F.2d 654 (D.C. Cir. 1989), *cert. denied*, 107 L. Ed. 2d 737 (1990).

Based on the foregoing, we affirm the Mass Media Bureau's conclusion that the material in question was not indecent. Accordingly, pursuant to Section 1.115 of the Commission's Rules (47 C.F.R. Section 1.115) your Request for Reconsideration of the October 26, 1989, determination, which the Bureau has referred to the Commission to be treated as an Application for Review, IS DENIED.

This letter was adopted by the Commission on January 24, 1991.

BY DIRECTION OF THE COMMISSION

Donna R. Searcy
Secretary

Commissioner Duggan dissenting and issuing a statement.
Attachment

February 8, 1989, 6:25 P.M.
All Things Considered # 890208 "GOTTI"

> **MV**: Male Voice
> **MS**: Mike Schuster
> **JG**: John Gotti
> **OV**: Other Voice

MV: He is known in New York as the dapper Don and the book published last year labeled him a Mob Star. John Gotti has become the nation's most notorious gangster. In recent years he's been acquitted in two celebrated trials and now he's facing new charges in connection with an attempted killing of a labor official. Through it all, John Gotti has become a familiar face on television and a feared presence on the streets of New York. NPR's Mike Schuster has a report and a warning: the following story contains some very rough language.

MS: This is an excerpt from a wiretap. One conversation Gotti had with an associate some years ago before heading the Gambino family. The tape has been played in court. Gotti is browbeating the associate for not returning his phone calls. The other man claims his wife didn't pass along Gotti's messages. Gotti's threats are extremely profane.

JG: (Unintelligible) fucking (unintelligible) you understand me?

OV: (Unintelligible)

JG: Listen, I called your fucking house five times yesterday. Now if you want (unintelligible) fuck (unintelligible). Now if you want to disregard my fucking phone calls I'll blow you and the fucking house up.

OV: I never disregarded anything.

JG: Are you, call your fucking wife or will you tell her.

OV: All right.

JG: This is not a fucking game I (unintelligible) how to reach me days and nights here, my fucking time is valuable.

OV: I know that.

JG: Now you get your fucking ass (unintelligible) and see me tomorrow.

OV: I'm going to be here all day tomorrow.

JG: Never mind all day tomorrow (unintelligible) if I hear anybody else calling you (unintelligible) I'll fucking kill you.

MS: As head of the Gambino family, Gotti controlled vast crime activities including gambling, loan sharking, labor racketeering and gasoline bootlegging. Author Gene Mustane says Gotti has worked hard during the past three years to consolidate his control and defend the mob from the onslaught of federal and state prosecutions that the Mafia suffered.

Dissent:

January 25, 1991

In Re: Application for Review of Bureau's Dismissal of Indecency

Complaint Filed Against National Public Radio's *All Things Considered*.

Between 1975 and 1987, the Commission took a somewhat limited approach to enforcing its prohibition against indecent broadcasts. It took no action unless the material involved the repeated use, for shock value, of words similar or identical to the so-called "seven dirty words" satirized in George Carlin's "Filthy Words" monologue—a performance involved in the Supreme Court's *Pacifica* decision. See *FCC v. Pacifica Foundation*, 438 U.S. 726 (1978).

Four years ago, the Commission found that this standard was unduly narrow. It concluded that the definition of indecent material set forth in *Pacifica* appropriately includes a broader range of material than the seven specific words at issue in that case. As we noted in *Pacifica Foundation, Inc.*, 2 FCC Rcd 2698, 2699 (1987), "[t]hose particular words are more correctly treated as examples of, rather than a definitive list of the kinds of words that, when used in a patently offensive manner as measured by contemporary community standards applicable for the broadcast medium, constitute indecency." The D.C. Circuit upheld this Commission action, finding that the FCC had "rationally determined that its former policy could yield anomalous, even arbitrary results." *Action for Children's Television v. FCC*, 852 F.2d 1332, 1338 (D.C. Cir. 1988). In broadening its definition of indecency, however, the Commission never abandoned its earlier standard. The seven dirty words remained a key determinant of what the Commission would find unacceptable. The Commission, by its 1987 clarification, merely expanded the standard to include a wider range of indecent material.

In this case, however, it appears that the Commission is veering away from its former standard. Bending over backwards, perhaps—because the broadcast in question was by National Public Radio, and because it was a newscast—the Commission suddenly appears willing to ignore the standard that in the past has guided its decisions on indecency. One stark fact remains, however: the broadcast featured, in the course of a few seconds, ten repetitions of the dirtiest of "the seven dirty words." The word in question is the one expletive that has traditionally been considered the most objectionable, the most forbidden, and the most patently offensive to civilized and cultivated people: the famous F-word. That word, in the past—and especially its deliberate, repeated, gratuitous use—has almost always been sufficient to justify a ruling of indecency by the FCC.

Can its use in the NPR broadcast now before us be defended as necessary to inform the audience and to establish the character of John Gotti, the man who spoke it? Perhaps. Just one airing of the word, however, would be sufficient to accomplish that purpose. Although I recognize the importance of context to indecency determinations, I consider that the deliberate and repeated use of this word fits precisely the meaning of the word gratuitous: unnecessary and unwarranted. And such deliberate and repeated use, in my judgment, however noble the intent of the broadcaster, seems to me to fit the definition of pandering: catering to low tastes.

In the sensitive and controverted field of indecency enforcement, every Commission decision, however small or close, has precedential value. I am concerned that the Commission's departure here from its usual standard, though well-intentioned, could open the floodgates to the repeated, gratuitous use of language that has historically and legally been considered indecent or obscene. By such decisions as this, the Commission may unwittingly encourage a plethora of "newscasts" that purvey patently objectionable material under the cover of journalistic legitimacy. Such a development would not only be a misfortune for our national culture; it would also contravene the intentions of Congress and the courts, who have never suggested that broadcasters be given carte blanche to incorporate indecent material into news or public affairs programming. Although the Commission has twice declined to adopt an express "news exemption" for indecency enforcement, it may have implicitly created such an exemption here (See Citizen's Complaint Against *Pacifica Foundation*, 59 FCC 2d 892 (1976); *Infinity Broadcasting Corp. of Pa.*, 3 FCC Rcd 930, 937 n.31 (1987)).

The Supreme Court has held that government can constitutionally regulate indecent broadcasts in the interest of protecting children. I interpret this as including the protection of children from the gratuitous use of patently offensive language even in bona fide news stories. And so I dissent, in this case, from the Commission's decision that the broadcast in question was not indecent.

In the aftermath of the Gotti broadcast and subsequent litigation, National Public Radio moved to "strict" new guidelines: "We are tightening up," *Newsday* reported. "We are more careful today than we were before. We think there is greater sensitivity and we work harder at making sure the language we use is acceptable," said William Buzenberg, vice president for news at NPR. "We have heard from a lot of stations and a lot of listeners," he said.[14]

If the management solution in such cases is to use an electronic "beep" to cover the profanity, one must ask what is lost in the edit? Is broadcasters' self-regulation less harmful of *First Amendment* principles than the heavy hand of government regulation? In the end, the government can make it more difficult than it is worth for a broadcast manager to accept risky programming.

Manager's Summary

Broadcasters' greatest concerns over listener and viewer complaints should be that FCC action might lead to a more difficult license renewal, in the long run. In the short run, however, broadcasters appear to face the threat of only relatively small fines where broadcast indecency complaints are determined to be actionable.

Notes

1. *Branton v. FCC*, 993 F.2d 906 (D.C. Cir. 1993), cert. den.
2. Ibid., p. 908: "Mr. Branton now petitions for judicial review of the agency's decision not to proceed against NPR."
3. Ibid., p. 912.
4. Ibid., p. 908.
5. Ibid., p. 909.
6. Ibid.

7. Ibid.
8. *Office of Communication of United Church of Christ v. FCC*, 359 F.2d 994, 1005–06 (D.C. Cir. 1966).
9. *Branton*, p. 910.
10. Ibid.
11. Ibid., p. 911.
12. Ibid.
13. Ibid.
14. Rita Ciolli, "Fit to Print?" *Newsday*, Pt. II, p. 73, 13 May 1993.

Chapter 8

The Social Construction of Howard Stern: Shock Jocks and Their Listeners

"I never worry about the people out there. . . . I feel no responsibility whatsoever. People say to me the media could educate, the media could enlighten people, and I know that's such a crock."
—*Howard Stern*, CBS This Morning, *October 1993*

"Good taste would likely have the same effect on Howard Stern that daylight has on Dracula. . . . Remember the kid in seventh grade who could blow milk out of his nose? Well, that's Howard Stern."
—Nightline, *December 1992*

"There is no speech so horrendous in content that it does not in principle serve our purposes."
—*Benno Schmidt, president, Yale University, 1986*

When the New York Libertarian Party nominated shock radio announcer Howard Stern for governor in 1994, it raised the eyebrows of conservative columnist William F. Buckley, Jr. Stern's support to restore capital punishment in the state emphasized the satire: "Stern gave a halfway serious talk designed to stress the libertarian elements of his own political beliefs."[1] In Albany, however, Stern's road show turned into anything but a typical convention speech: "Joining the entourage," Buckley noted, "was a scantily clad woman with large breasts and a lavishly tattooed lesbian who claims to have had sex with space aliens."

By August, Stern had quit the race after a court ruled against him in a challenge to financial disclosure laws.[2] "I'm getting 15 to 20 percent of the vote," he complained. "It's a joke. I am not a candidate for governor. You just have to live with a governor who tells you how much stock he has in IBM but won't tell you what's in his head or his heart."[3]

Stern's media image has brought him fame and fortune, and it threatened to bring him political power. His case highlights the point that broadcast indecency is a form of speech that draws significant public attention. It affects the form and content of public communication.

The Media Icon of a Shock Jock

Howard Stern's media image was formed in the early 1980s—a time when he emerged as a ratings superstar.

Howard Stern

USA Today's first-ever best-seller list placed Howard Stern's autobiography *Private Parts* in the top spot. Stern became an American phenomenon when the Federal Communications Commission targeted his shock jock radio format with a series of indecency fines beginning in 1987.

During the 1980s Stern rose to the top of the ratings wars in the nation's largest radio markets, and his success led to a proliferation of imitators. Stern, a communication graduate of Boston University, began his career as a conventional radio deejay in 1976. At WCCC Hartford in 1978 he experimented with a mix of outrageous telephone talk and music. To protest gas lines he urged a protest called "To Hell With Shell." He developed Dial-A-Date, a sort of lurid take-off on television's *The Dating Game*.

His success landed him jobs in Detroit and Washington, D.C., where newscaster Robin Quivers became his female sidekick. After tripling station ratings, he left for WNBC New York, where his caustic brand of humor and promotion of negative racial stereotypes earned him critics and more financial success. Stern, the forty-something son of a radio engineer in Long Island, found himself "wanting to be black" in seventh grade as his white suburban neighborhood integrated. The Sterns eventually moved to the all-white community of Rockville Centre, where Jews were excluded from the country club.

Quivers is a black steelworker's daughter who grew up in a Jewish neighborhood in Baltimore. She has defended Stern's usage of black dialect to respond to callers. Stern's ratings at WXRK-FM New York are highest with men eighteen to fifty-four years-old. Commercials on the flagship stations reportedly sold for as much as $2,000, the highest in the market in 1992. Parent company Infinity Broadcasting was fined $500,000 in 1993 for alleged indecency on the *Howard Stern Show*, a program distributed in morning drive time coast to coast. At one point, the FCC threatened but retreated from blocking Infinity's $100 million purchase of stations in Atlanta, Boston, and Chicago because of Stern.

Stern's program was taken off the air by Chicago station WLUP following a series of listener letters to the FCC. One listener complained: "There was a jovial discussion about cutting off a woman's legs and beating her with them, and then cutting off her breasts, putting them in a blender and making her drink the results." Stern responded with a $45 million breach of contract lawsuit against Evergreen Media, which also has been fined by the FCC for its *Steve and Gary Show*.

The FCC actions against Stern have centered most frequently on his violation of an indecency statute by making explicit references to sexual behavior. His 1988 "Christmas Party" broadcast, for example, featured a man playing the piano with his penis.

Stern was attacked as insensitive when in 1982, after an Air Florida plane plunged into the Potomac, he called to ask what a one-way fare from National Airport to the 14th Street Bridge cost. He defended the action as satire intended to produce social action.

Stern's commercial success can be seen as part of the larger cultural trend in all forms of media toward explicit references to sexual behavior, violence, and blunt attacks on social institutions.

Sources: *People* (22 October 1984); *Broadcasting* (16 August 1993); *USA Today* (28 October 1993); and FCC documents (26 October 1989 and 27 January 1993).

The Social Construction of Reality

Sociologists Peter L. Berger and Thomas Luckmann's *The Social Construction of Reality: A Treatise in the Sociology of Knowledge* (1967) is a starting point for understanding how and why media performers such as Howard Stern have become so popular.[4] The contention is that society engages in a human communication process of establishing certain knowledge" as "reality."[5] In other words, we collectively decide "what is." In the case of broadcast indecency, our society judges and decides that certain sexual language is objectionable because cultural norms are violated:

> Every culture has a distinctive sexual configuration, with its own specialized patterns of sexual conduct and its own "anthropological" assumptions in the sexual area. The empirical relativity of these configurations, their immense variety and luxurious inventiveness, indicate that they are the product of man's own socio-cultural formations rather than of a biologically fixed human nature.[6]

Inasmuch as a performer such as Howard Stern is able to communicate through mass media to large numbers of audience members, he has power to take part in defining social reality—a power that can be treated as an ideology of sorts.[7] We apparently need such performers in modern society to help reflect changing mores as well as to participate in the evolution of them. In the absence of traditional figureheads, performers such as Stern preside over the norm definitions. As we will see in this chapter, the news media's coverage of broadcast indecency by a variety of performers also is a social construction that defines the issue.

The Construction of Indecency

Whether we are talking about Howard Stern or other media performers, government action against broadcast indecency involves an attempt to define norms. The difficulty is that Federal Communications Commission action will not always parallel individual or community norms. The examples are many. According to Dennis Wharton of *Daily Variety*, Group W's KYW-TV, Philadelphia, came under FCC scrutiny for broadcasting a program about "crouch dancing" at a bar:

> Complaint involves a November 1991 segment of *Jane Wallace Live*, in which the program's host investigated a south New Jersey bar specializing in "crouch dancing," in which nude women gyrate close to the faces of male patrons.
>
> The program featured call-in comments from listeners, many of whom spoke in graphic sexual terms. The Federal Communications Commission recently fired off a letter to KYW-TV claiming the pro-

gram may have violated indecency rules because it was aired at 10 A.M., a time "when there is a reasonable risk that children may have been in the audience."

Jane Wallace Live was a program developed by Group W in a bid to crack the national syndication market. The show proved short-lived, however, folding after only several months on KYW.

Group W Vice President Gil Schwartz defended the show on First Amendment grounds. "He said the program 'succeeded in uncovering and reporting on allegations of serious misconduct at a local neighborhood bar.' "[8]

Radio personality Howard Stern has become the lightning rod in media coverage of the indecency issue. As seen in this 1993 *USA Weekend* story, many Americans would "muzzle" him:

The station admitted last month that it had begun to edit the raciest parts of Stern's program, explaining that it wished "to avoid further complaints" while the FCC fine is pending.

A sampling of reader mail:

> Please take sickos like Stern off the air. The Founding Fathers did not have filthy mouths in mind when they wrote the First Amendment.
> —JESSIE WOLF, Clifton, Colo.

> I am a 57-year-old who finds Stern intelligent, insightful and usually extremely funny—which is a lot more than can be said for Terry Rakolta, the Michigan "decency" crusader, who seems too lazy to monitor what her kids listen to and would like the FCC to do it for her! If you're concerned about lewd material, I suggest you watch the daytime talk shows during the ratings sweeps. They trot out every perversion known to man. If you don't like Stern, change the station!
> —WILLIAM C. NICOLL, JR., Vineland, N.J.

> The intention of the First Amendment was to allow us to speak out against our government without fear of retribution—not to slander groups of people or individuals. Stern never will know what it is like to continually struggle for

recognition, as do minorities, gays, and women. Every time someone gets on the radio and ridicules women, it reinforces nonsensical notions further in listeners' minds.

—TRACEY BATES, Fredonia, N.Y.

We must ask ourselves whether we have the right to censor a person because we don't like what he says. We do not. But we do have the option of changing the station. Much of Stern's attraction comes from being shocking. The novelty will wear off, and his audience will dwindle. Until then, he should be limited to late-night broadcasting. Parents cannot control everything in their children's lives; society should share responsibility. Most people would agree we don't want future generations of nothing but Sterns. One already is too many.

—KERRY A. MCGRATH, Bellingham, Wash.

How much filth are we going to allow before it's too late to turn the tide in favor of decency? We have given up a spiritual foundation for the toilet bowl of entertainment.

—NEWTON E. BUNCE, Redlands, Calif.

Jonathan Alter's disingenuous, self-righteous article is of greater danger to America than any vulgar pop celebrity. Do Alter's dubious credentials as an editor of a national magazine (*Newsweek*) qualify him to be a sanctimonious judge for the entire country? Absolutely not. The First Amendment gives all Americans the right to exercise that judgment personally.

—JIM FRITZ, Aurora, Colo.

I respect the right to express an opinion, but I also believe the opinion should be expressed responsibly. I have the right to drive a car, but does that mean I can drive all over your lawn? I have the right to own a gun, but does that make it OK for me to shoot people who offend me? There always will be those like Mr. Stern who will push the limits of good judgment.

—MICHAEL GILHULY, Clinton Township, Mich.

How You Voted

Yes, get shock jocks like Stern off the air	65%
No, let them alone	25%
No, but confine them to late night	10%[9]

The social criticism of radio personalities such as Howard Stern, and the move to clean up the airwaves, might best be seen in Chicago, where a local newspaper called Stern radio's "bad boy" and "the industry's most talked-about air personality."

> While Stern's number seventeen overall ranking in Chicago may be underwhelming (he ranks thirteenth among listeners age twenty-five to fifty-four), some say it's still too soon to count him out.
>
> "It's still a little early," said Steve Butler, president of *Inside Radio*, a daily industry newsletter."[10]

The *Minneapolis Star Tribune* included this Howard Stern item under the "News of the Weird" section:

> Acting Federal Communications Commission chairman James Quello, reacting to radio shock-jock Howard Stern's statement that Stern would have to answer to a "higher authority" than the FCC for his so-called indecency, said, "I wouldn't be a bit surprised if some-day a lightning bolt comes out of the sky and hits (Stern) right in the crotch."[11]

That there are more serious issues at stake, consider the Thomas Jefferson Center for the Protection of Free Expression. This organization issued a "muzzle award" in 1993 to the former FCC Chairman:

> Former FCC Chairman Alfred Sikes earned a 1993 Jefferson Muzzle for the effort to muzzle radio personality Howard Stern by assessing a record $600,000 fine on Infinity Broadcasting Corporation for airing Stern's program. The center believes the FCC indecency action fell far from what courts would require to satisfy First Amendment rights in obscenity cases.
>
> "This award is not a judgment on Mr. Stern or the language he uses on the air; that is for the public to decide," O'Neil said. "It is a judgment about governmental censorship—an action the Founding Fathers considered more dangerous than even the most obnoxious ideas."[12]

Interim FCC Chairman James Quello in 1993 told lawmakers on Capitol Hill the commission could initiate revocation proceedings against licensee Infinity for Stern's repeated violation of the law:

"There's no question that we would go the extra final step," said Quello. He also told the members there may be enough evidence of an Infinity indecency record to begin revocation hearings, should the commission cite Infinity again and decide to proceed in that direction.[13]

People magazine, in its construction of Stern, has also written about his "side-kick" Robin Quivers. "We're like a fungus," says Quivers. . . . "The FCC does what it does," she says. "There are certain people in the country who don't understand the Constitution. Howard and I are the most misunderstood people in broadcasting:"

She studied nursing at the University of Maryland, but as a nurse she found herself suffering the profession's classic complaints—"bad hours, low pay, and lack of respect." After Quivers spent two years in the Air Force, a stint with a broadcast consulting firm in 1978 pushed her into a radio career.[14]

Listener Al Westcott of Las Vegas has supplied the FCC's mass media bureau with tapes documenting Stern's broadcasts on affiliate KLSX-FM, Los Angeles. In 1992, the FCC cited Stern's objectionable language in passing references:

- "The closest I came to making love to a black woman was masturbating to a picture of Aunt Jemima."
- Stern suggested that actor Paul Reubens (Pee-Wee Herman), who was charged with exposure in a Florida porn theater, should ejaculate on a TV audience "and really give it to them right in the face."[15]

According to the *Washington Post*, sixty hours of tapes were reviewed: "In his Feb. 2 letter to the FCC, Al Westcott, of Las Vegas, cited 'discussions concerning the size of Mr. Stern's penis,' 'in-studio female guests who disrobe and allow Mr. Stern to spank them,' 'discussions concerning Ms. [Robin] Quivers's [Stern's sidekick] stated affection for sodomy," and other topics discussed by Stern that alarmed him when he was in L.A."[16]

The trade magazine *Broadcasting* (now *Broadcasting & Cable*) found it needed to issue a special statement to report on complaints against Stern: "Readers may find some language in the following story offensive. It goes beyond what the editors would ordinarily admit, but—in *Broadcasting*'s role as the book of record—is published to inform readers about actual language or material that has warranted challenge or sanction by the FCC."[17]

Stern's show went on the air on Los Angeles's KSLX-FM (97.1) in July 1991, and surged to number one in the nation's second largest market by summer 1992. The *Los Angeles Times* found that Stern's success might lead to copycats. Programming consultant Dan O'Day was quoted as saying imitators would be "loud and opinionated and obnoxious and dirty."[18] The article described Stern's content as unusual: "Stern's show focuses heavily on his sex life—he discusses masturbation and love-making with his wife—and his sexual fantasies. He regularly baits in-studio guest actresses, strippers, and other women to disrobe in front of him, or to submit to a spanking, often labeled 'butt bongo.' He asks celebrity guests about their sex lives. 'Lesbian Dial-a-Date' is a regular feature. Stern's bathroom habits are another favorite topic." O'Day said of the content: "Howard's interests are on the junior high school level." O'Day said, "He really does read *Penthouse*, he really does rent Kung Fu movies. He really doesn't like art movies. It works for Howard because he's genuinely interested in that. Others would do it only because it works for Howard Stern and they'd be ripping it off."

Jeffrey Yorke of the *Washington Post* said the popularity of "risque radio" soared as announcers such as Stern tested the limits of content. Doug "Greaseman" Tracht's show was about as popular as Stern's:

> On one show, the Greaseman, known nationally for creating stories that rely largely on double entendre and innuendo, told listeners the tale of a young man who, while a customer in a butcher shop, encounters an attractive female, the butcher's apprentice. She makes a suggestive reference to sausage.[19]

"Anyone who happens to tune in to this program will get a barrage of this sort of 'entertainment' within 10 or 15 minutes," Joyce M. Zuckerman said in her 1991 complaint, according to Yorke.

Why are announcers such as Greaseman and Stern so popular? One Los Angeles Times story quoted avid Stern listener Frank LaVeaga who saw the appeal as "kind of a '90s thing where everybody's growing up but they don't want to let go of their adolescence." Added Stern fan Donnie Gallagher: "He says what everyone else wants to say—but is afraid to."

Topless and Shock Radio

Howard Stern's style has its roots in radio of the early 1970s. The shock radio format, particularly when it comes to indecency, is closely related to what the media once called "topless radio."

Topless Radio

Radio announcer Bill Ballance called his 1970 KGBS, Los Angeles, show *Feminine Forum*. Women were encouraged to reveal their secrets—particularly sexual secrets.

Ballance has told a San Diego newspaper that ratings success was noticed on the East Coast: "I was interviewed by the *Wall Street Journal, Newsweek, Time*. And some yellow journalist scum (a reporter from the *New York Daily News*) nicknamed my show 'topless radio,' which of course, it wasn't." What it was, of course, was American mass media reacting to and participating in the sexual revolution of the 1960s.

The Nixon-era Federal Communications Commission, presided over by Chairman Dean Burch (1969–74), attacked the format. In a speech to the national Association of Broadcasters meeting, Burch called such shows "electronic voyeurism."

A 1973 case against WGLD, Oak Park, Illinois's *Femme Forum* followed discussion of "How to keep your sex life alive?" One female listener responded that the answer was her craving for peanut butter: "I used to spread this on my husband's privates and after awhile, I mean, I didn't even need the peanut butter anymore." Dissenting FCC Commissioner Nicholas Johnson called the resulting $2,000 FCC fine "arbitrary, unwise, and unconstitutional." A 1975 appeals court decision upheld the right of the FCC to act against broadcast indecency.

Sources: the *San Diego Union-Tribune* (16 March 1993); the *New York Times* (14 April 1973); the *Los Angeles Times* (19 August 1987); and *Illinois Citizens Committee for Broadcast v. FCC*, 55 F.2d 397 (1975).

The Political Significance

That an announcer such as Howard Stern could be even considered for high political office in New York State suggests the obvious political implication of broadcast indecency. As an attention-getting vehicle, provocative speech may be used to secure actual political power. At a more subtle level, however, such public usage of previously socially objectionable language may lead to social change. By changing the form of political and nonpolitical (if there is such an animal) speech, one also changes content. Social theorists might try to distinguish between economic and political resources and power as entities of the "society," and those cultural forces affecting meanings and practices.[20] However, Stern's ability to affect cultural meanings and social practices—particularly communication practices—cannot be completely isolated from economic and political issues. This is especially the case when society economically, and perhaps even politically, rewards such personalities for the behavior the government is attempting to sanction.

Stern's Rhetoric and Its Political Uses

A cursory review of Stern's political stands would seem to place him as populist, extremely libertarian on speech issues, exhibiting a change-oriented approach to government, yet quite traditional on issues such as crime. Future researchers need to employ systematic content analyses to make further assessments of the political significance of his speech.

Why Do They Say It on the Air?

The obvious answer to why personalities such as Stern do it is that their programs continue to be popular. As ratings builders, these shows produce profit. There is also historical context of the post-*Pacifica* era. Dominick, Sherman, and Copeland (1990) argued that lack of FCC action following the decision "may have encouraged broadcasters to push the limits a little too aggressively."[21] It may be a simple case of economic reward and lack of legal sanction that has promoted expansion of the boundaries of broadcast speech.

Who Listens to Howard Stern?

In 1990 WXRX radio submitted evidence to the FCC—a Gallup listenership study of 250 heads of households where children ages six to eleven resided. Thirty-five percent (N = 87) of respondents (N = 252) said their children listened to New York market radio stations between six and ten in the morning; but only one respondent said their child listened to *The Howard Stern Show*—and that was with parental supervision.[22] Only six percent of these parents said they listened to the shock jock.[23]

A Real Threat?

The argument, of course, is that Stern's program is not a possible threat to the well-being of children. The FCC bases its regulatory rationale on the need to protect children who might be in the audience. Infinity's research data seem to support two conclusions:

1) Most young children are not unsupervised audience members for shock radio programming.
2) The evidence is inconclusive that a child hearing indecency will be harmed.

Howard Stern, saying "free speech is the single greatest freedom we have in this country," told *Playboy* magazine in 1994 that there should be no limitations: "The rule that I follow is this: I should say anything that I think is funny."[24] He rejects the notion that what he says is bad:

> What is this bugaboo about sex? I mean, what is this hang-up that we have? To me a penis is like your arm. You know? Your penis or your vagina is just another part of your body. But as adults we're so freaked out by it that we almost laugh at it—we find it funny to say. We're so fucking uptight. We are fucking crazy. We've gone mad.[25]

What is Stern's definition of indecency? He says he does not define it:

> If you're asking me subjectively to make a call on what indecency is, I would have to go with child pornography, abuse, murder, raping young boys if you're a priest. All these are indecent. But as far as talking about them on radio, I have no problem.[26]

Stern argues that the FCC has practiced "selective enforcement" by distinguishing between medical and comic discussion of sex:

The laws have to apply to everyone. Still, if you have a psychiatrist sitting up on stage with [talk-show host Phil] Donahue, then it's legitimate. If you have me sitting around cracking jokes about sexuality, that's considered frivolous. To me, free speech is free speech. You can't decide in what context you can speak about these things.[27]

The FCC, however, argues that every indecency complaint must be considered in context to determine if the content is "patently offensive."

What Next?

Infinity's $1.7 million settlement only cooled Stern's heat temporarily. Just one month later, in October 1995, word came that a hispanic group wanted revokation of the license of KSLX-FM (Greater Media), Los Angeles.[28] In April, Stern had commented on the shooting death of pop star Selena. With her music and gunshot sound effects playing in the background, Stern's comments went "far beyond the boundaries of contemporary community standards," the National Hispanic Media Coalition said.[29] Stern's "apology," which was delivered in Spanish, appeared to anger the group further.

Stern's brash approach continued to affect his distribution network. In Chicago, for example, where Stern has filed a breach of contract lawsuit against Evergreen's WLUP, a second station—Cox's WCKG-FM—dropped the program following an advertiser boycott. The station's manager said: "It was the first time in my life I'd ever seen anybody do that."[30]

It remains to be seen whether Stern's in-your-face approach, which sometimes includes comments about former business associates in the broadcast industry and FCC commissioners, will continue to be tolerated by broadcast regulators. In a complaints-driven system, the FCC must continue to be responsive to the steady stream of audience complaints about Stern's remarks. Stern, however, continues to be profitable because of marketplace interest. His second autobiography has been published and like the first climbed to the top of the charts. This popular media icon continues to help define the FCC's struggle with marketplace principles and decency laws.

Manager's Summary

1. Broadcast managers who carry Howard Stern's show are likely to spark a lot of commentaries in their communities. They must be mindful of local community standards and the potential for complaints.
2. The future for Stern and other shock jocks is likely to depend upon two factors: (a) the continued market for such programming, and (b) the willingness of the FCC to respond to audience complaints.

Notes

1. William F. Buckley, Jr., Universal Press Syndicate, "Howard Stern stumps as Libertarian," *Omaha World-Herald*, 1 June 1994, p. 11.
2. Associated Press, "Stern quits race in N.Y. to keep finances a secret," *Omaha World-Herald*, 4 August 1994, p. 46.
3. Ibid. The wire story said Stern's "raunchy on-air stunts have brought his syndicated radio show millions of fans. . . ."
4. Peter L. Berger and Thomas Luckmann, *The Social Construction of Reality: A Treatise in the Sociology of Knowledge*. New York: Anchor, Doubleday, 1967.
5. Ibid., p. 3.
6. Ibid., pp. 49–50.
7. Ibid., p. 123.
8. Dennis Wharton, *Daily Variety*, 28 April 1993, p. 3.
9. *USA Weekend*, 25 April 1993, p. 14.
10. Dan Kening, "Stern's shock-talk show not playing in Chicago," *Chicago Tribune*, (zone n), 21 April 1993, p. 1.
11. Chuck Shepherd, *Minneapolis Star Tribune*, 15 April 1993, Metro Edition, p. 2E.
12. *PR Newswire*, 12 April 1993.
13. Bill Holland, "Infinity license in danger of revocation over Stern?" *Billboard*, 10 April 1993, p. 69.
14. *People*, 5 April 1993, p. 103.
15. Jeffrey Yourke, "Stern talk results in fine," *Washington Post*, 27 October 1992, p. C7.
16. Ibid.
17. Joe Flint, "Hounding Howard: FCC's $100K fine, Radio Personality Howard Stern," *Broadcasting & Cable*, Vol. 122(44), 26 October 1992, p. 8.
18. Claudia Puig, "Howard Stern: The next generation of talk radio; industry insiders predict that the nationwide popularity of the renegade morning man will spawn a host of imitators," *Los Angeles Times*, 8 October 1992, p. F1.
19. Jeffrey Yourke, "Locking on the shock jocks," *Washington Post*, 18 August 1992, p. D7.
20. Denis McQuail, *Mass Communication Theory: An Introduction*, 3rd ed. London: Sage, 1994, p. 61.

21. Joseph Dominick, Barry L. Sherman, and Gary Copeland, *Broadcasting/Cable and Beyond: An Introduction to Modern Electronic Media.* New York: McGraw-Hill, 1990, p. 416.
22. Gale D. Muller, "WXRK radio special listenership study," The Gallup Organization, Princeton, NJ, December 1989.
23. Ibid.
24. "*Playboy* interview: Howard Stern, a candid conversation with radio's best-selling worst nightmare about his small penis, big plans, blue language, and red-hot career," *Playboy*, 41(4):55–68, 158–160 (April 1994).
25. Ibid., p. 66.
26. Ibid., p. 67.
27. Ibid.
28. Chris McConnell, "License Revokation Sought over Stern's Remarks," *Broadcasting & Cable*, 2 October 1995, p. 23.
29. Ibid.
30. Donna Petrozzello, "Stern Loses FM in Chicago; Moves to AM," *Broadcasting & Cable*, 9 October 1995, p. 37.

Chapter 9

The Question of Effects from Indecent Broadcasts

Regulation of broadcast indecency rests on the assumption that exposing children to foul language is harmful. The government, through the Federal Communications Commission, uses the assumption to argue that it must assist parents in the job of protecting their kids. The effects issue, however, is anything but settled by academic research.

The Problems with Measuring Mass Communication Effects

Years of social research suggests that the effects from mass media messages can range from very small to very large. Severin and Tankard (1992) conclude that "most mass media effects do not occur 'across the board,' but are contingent on other variables."[1] So, if our concern is whether broadcast indecency has a negative effect on children, the answer of mass communication would be, "It depends." It depends on which children and under what circumstances there are exposures to certain messages, what the messages are, and to what extent those messages conflict with or reinforce other messages the children receive.

In the area of television violence, for example, one estimate was that an "average child" watches more than 100,000 violent episodes and observes more than 13,000 deaths by age twelve.[2] The current findings, however, support the notion that watching TV violence is a cause of some aggression, but the effect might be quite small.[3]

A Question of Social Influence

The issue can be framed in a question: To what extent are mass media messages influential in the "socialization" of children and adults?[4] In the case of child socialization, we most often see references to "modeling" or "social learning theory." The idea is that individuals learn new behavior by observing others—whether it be in person or through mass media messages.[5] Social learning theory, as one that focuses on the individual, falls short of explaining the influence of mass media in shaping cultural change, such as the acceptance of profanity in society.

We can see that the question of profanity is a broad one affecting more than broadcast speech in this country. In 1994, *Omaha World-Herald* Editor G. Woodson Howe argued that, "The vulgarization of language in public forums has accompanied a new, more strident tone, once rarely heard in public debate."[6] Howe wrote that mainstream news stories now use words (screw, sucks, pissed, crap) considered "dirty" a generation ago. Broadcaster and entertainer popularity, he reasoned, seemed to be assisted by profanity. As concerned as the editor was, he recognized the First Amendment problem: "This is not to argue for Federal Communications Commission crackdown on television content. Nor is it a call for a return to the Motion Picture Production Code that limited artistic expression in the 1960s. Newspaper editors have too much reverence for the First Amendment to suggest a government role in determining taste."

He added: "But there should be room for self-regulation in the mass media." That conclusion lead the newspaper to edit columnists and sports interviews to eliminate crude words not just in the interest of children: "We do it also because thousands of adult readers are offended by gutter language that they themselves don't use in public."

The Special Issue of the Effects of the Broadcasts on Children

An exhaustive review of research on the effects of broadcast indecency in 1992 suggested that "the available evidence does not justify the 24-hour ban or, for that matter, even the older more limited restriction on indecent materials."[7]

Key Questions

Academic researchers Donnerstein, Wilson, and Linz have posed three research questions:

1. What are the effects of exposure to indecent language?
2. What are the effects of exposure to song lyrics or poems that contain sexually explicit language or sexual references?
3. What are the effects of exposure to movies that include nudity and scenes depicting sexual matters?

The researchers could find no studies on the effect of children's exposure to indecent language, some evidence that children do not understand indecent song lyrics, and evidence that exposure to indecent images has a limited effect.[8]

Harmful Effects

The question, then, turns to the issue of "harmful" effect, which has two elements: "one based on a societal judgment about what constitutes harm, and the other based on the reliability and validity of scientific evidence necessary to demonstrate such harm."[9] The researchers contend that they cannot answer the normative social question about harm—that, they say, deals solely with philosophy and politics. On the second, social science, question—the one where evidence is slim—the researchers say there are limitations in establishing causal connections.

The conclusion, to review, is that children under twelve years old have limited abilities to understand sexual references used in language. Older children, thirteen to seventeen, may understand the content, but "they are likely to have developed moral standards which, like adults, enable them to deal with broadcast content more critically."[10] This last point could be seen as more of an open question. It may well be, after all, that there are sizable numbers of adolescents who have not developed such "moral standards" and are susceptible to what society could define as "harmful effects."

One nonbroadcast example of this issue is found in the concerns raised by some adults about the content of music lyrics. We now turn to an example from Omaha, Nebraska.

Rap Music and the Omaha 1992 Case Study

In 1992, Luther Campbell of the group 2 Live Crew visited an Omaha record store fighting obscenity charges for selling the group's recordings. The case was settled after stores agreed to stop selling the music.[11] A city council member (since defeated for re-election) and a group calling itself Omaha for Decency called lyrics in the song "Sports Weekend" (see box below) obscene. The city council member, calling a variety of media unfit, said: "I think we have a moral obligation, as legislators and as parents, to set down guidelines for our children."[12] Record-store owners in Omaha claimed they had not intended to sell the music to minors. One parent of a teenager, Meg Graves, told the Omaha newspaper that censorship isn't right: "As parents, everybody has to watch their own kids and teach their own morals."[13]

Creighton University Law Professor G. Michael Fenner, who teaches constitutional law, wrote that one cannot go back and stop offensive speech:

> They cannot stop Luther Campbell, lead rapper of 2 Live Crew, any more than they can stop Larry Flynt, whose Hustler magazine leaves nothing to the imagination, any more than they can stop Andrew Dice Clay, whose "comedy" reportedly includes the one about the man who has a right to have sex with his daughter because he pays her tuition, any more than they can stop Howard Stern, whose radio program includes blasphemy, the audio portion of what purported to be a live sex act performed next to a phone. . . .[14]

Fenner reminded readers that "scapegoatism" occurs when people stop the message about problems instead of dealing with the problems. In 1940, he said the FCC deleted a verse from Woody Guthrie's "This Land is Your Land" because of vague references to communism ("That side was made for you and me").[15] Subjective notions on decency, based on presumed antisocial effects of exposure to indecency, ultimately return us to issues of social psychology and sexual norms.

Sexual Dysfunctions

The field of abnormal psychology emphasizes biological and psychosocial causes in the study of sexual dysfunction in adulthood.[16]

While one might suspect that childhood exposure to sexual material might produce negative sexual behavioral effects later, mass media fall within a group of factors, including "sociocultural, personal, and interpersonal matters."[17]

Under psychosocial causes, sexual dysfunction can occur from lack of exposure to sexual matters—a condition that can lead to unhealthy inhibitions. Society can lead an individual to think of sex as "dirty" based upon prohibitions of certain conduct: Society can also foster ignorance and myths about standards of sexual performance and ability. Many sexually dysfunctional people are simply ignorant about sexual anatomy and physiology and harbor erroneous beliefs about what constitutes satisfactory sexual functioning.[18]

For those who would fear childhood exposure to sexual language, at least in the case of sexual problems, most "are not the result of traumatic experiences in childhood."[19]

Thus, it is fair to say that it is not talk of sexual matters that is inherently bad for young or old people. The more important issue is the accurate, scientifically valid portrayal of sexual matters. The decision to prohibit discussion of sex could be more harmful than the current situation, where humorists use stereotypical views of sex as devices in humor. Instead of banning indecency from the airwaves, concerned adults might want to focus their efforts on requiring responsible discussion of sexual matters.

We seem to be at cross purposes when we attempt to regulate the marketplace of free speech. For one, attempts to stop such speech often create publicity that makes it more popular. For another, such "bad" speech, if popular, would seem to have some value. Otherwise, why would listeners, viewers, readers, and advertisers pay to support it? We turn to the issue of money and indecency in the next chapter.

Sports Weekend (Nasty As They Want To Be Part II)
2 Live Crew

Song Name	A--	B-tch	Di-k	God Da--	F---	F--- That Sh--	Mother F--er	Ni--er	Pu--y (Pump That)	Ti--ies (Whore)	S--t	Anal Inter-course	Oral Inter-course
Pop that Pu--y	1	4	2		3	11	2		20		1		
A F--- is a F---		8	2		43		5	15	2	(1)	3		4
Just A Little More Head		2	4		1								
Ugly as a F---	1	8	1		9			3		1	2		
Fraternity Joint	3	1		1	4		8	1			2		
Here I Come	2				1		4	4		1	2		
I Like It, I Love It				8							2		
Mega Mix V											2		35
Freaky Behavior	1	2	1		7		2	2	1	(1)	2		3
12" Long		4			2								
Ain't No Pu--y Like Some Hot Head		3	4		3				80		20		5
I Ain't Bulls---tin III		5	2		5		5				27		3
The Pu--y Caper	6			5	3		3		2		5	8	
Up a Girl's A---	1												
Who's F---ing Who		2			25				1		1		
Pu--y for Those Who Like to F---													
(Instrumental Only)													
Totals	15	39	16	14	106	11	29	25	100	5	67	8	50

Other Abusive Messages:
1. You would f--- Satan for the righteous dollar.
2. One song actually is one male and one female "challenging" each other in a "game" or anal intercourse.
3. Another song boasts of the fun in group anal intercourse.
4. Have a much sex as possible with anyone, anytime—always use a condom and everything will be straight.
5. Making lots of money is all I know.
6. She's not like a woman, she's just a f---.
7. No woman wants a commitment to a man—she just wants to f---.
8. Nobody can f--- with us in this rap game.

FIGURE 9.1   Omaha for Decency Coding Sheet of "Sports Weekend"

Sports Weekend (Nasty As They Want To Be Part II)
2 Live Crew

A Sick Poem:
I f—-ed an old gal in a graveyard. God da— her old soul she was dead.
The maggots crawled out of her a--hole when I finished my job in there.
Her hair slipped off on her head and I seen I'd committed a sin.
So I pulled a straw out of my pocket and sucked out the load I shot in. (Followed by vile laughter.)

FIGURE 9.1   *Continued*

Notes

1. Werner J. Severin and James W. Tankard, Jr. *Communication Theories: Origins, Methods, and Uses in the Mass Media*, 3rd ed. New York: Longman, 1992, p. 265, citing Chaffee, 1977.
2. Ibid., p. 255.
3. Ibid., p. 258.
4. Melvin L. DeFleur and Sandra Ball-Rokeach, *Theories of Mass Communication*, 5th ed. New York: Longman, 1989, p. 209.
5. Ibid., p. 212.
6. G. Woodson Howe, "Coarsening of language widespread," *Omaha World-Herald*, 25 September 1994, 2A.
7. Edward Donnerstein, Barbara Wilson, and Daniel Linz, "Standpoint: On the Regulation of Broadcast Indecency to Protect Children," *Journal of Broadcasting & Electronic Media* 36(1):111–117 (Winter 1992).
8. Ibid., pp. 111–114.
9. Ibid., p. 114.
10. Ibid., p. 116.
11. John Taylor, "The music of anger and outrage: Is rap dangerous, important, or just annoying? It depends on who's talking," *Omaha World-Herald*, 12 July 1992, E1–2.
12. Joy Powell, "Exon adds TV, movies to list of objections," *Omaha World-Herald*, 29 April 1992, p. 9.
13. Supra, note 11 at E2.
14. G. Michael Fenner, "Address problem, not those who rap about it," Another point of view, *Omaha World-Herald*, 24 April 1992, p. 23.
15. Ibid.
16. Richard S. Perrotto and Joseph Culkin, *Exploring Abnormal Psychology*. New York: HarperCollins, 1993, pp. 172–173.
17. Ibid., p. 173.
18. Ibid.
19. Ibid., p.174.

Chapter 10

Making Money: Advertising and the Issue of Broadcast Indecency

Commercial broadcasters, of course, are in the business to make money. This is done by selling advertising. Advertising is sold by delivering audiences/consumers to advertisers through the use of highly rated broadcast programs.

The Profit Motive: Money from Ratings

Performers such as Howard Stern are able to show they have audiences/consumers listening through the evidence of station ratings. The larger the estimated audiences, the larger the ratings and the higher the prices for advertising spots.[1] Broadcasting reaches "all family members" at home and away.[2] In the area of television advertising targeted at children, the Children's Television Act of 1990 attempted to provide for minimal protections and urged self-regulation by broadcasters.

In the case of radio, declining profit margins in the past two decades may be helping fuel the drive to attract audiences through shock radio:

> More than half of all radio stations in the United States operate at a
> loss. . . . Nearly 300 radio stations have "gone dark" (closed down),
> more than half of them alone in 1991. Especially dramatic evidence
> of radio's declining value came in 1990 when MTV offered a Georgia
> AM station as first prize in one of its promotional contests.[3]

Stations in the largest markets, where broadcast indecency might be expected to be accepted as meeting community standards, do seem to be most profitable. The largest markets, according to the National Association of Broadcasters, averaged profit margins above 25 percent in the early 1990s.

When Advertising and Indecency Collide

There is a temptation for stations to accept most any advertisement in the name of revenue. This can be seen in terms of wants and needs: "Owners want the preservation of the firm and its assets, high rates of return on their investments, company growth, and increase in value of the firm and, thus, their investment."[4] Because of target marketing and the role of advertising agencies in purchasing, however, the competition for audience does not translate into a same level of competition for advertising: "Competition for advertising revenue . . . is significantly limited except among media with similar qualities and content delivery forms."[5]

At the same time, however, audiences may judge media performance on the public interest standard. Social critics of sexual references in advertising make interpretations: "Like violence, it is too often assumed to be a literal form of social control."[6] Sexuality can be seen by researchers as either oppressive (racist or sexist) or liberating.[7] Marketplace pressures, or even attempts at outright censorship, "must deal with the variables of American society."[8] In other words, management decisions about particular advertising must be seen within the social context of the moment. It would be unfair to say that most broadcast managers do not consider the pressures of their audience, the local community and its politics, as well as potential economic consequences from a decision to air or not air a controversial advertisement.

Abortion Advertising

For several years local television stations have struggled with the issue of the broadcast of graphic antiabortion advertising. By 1992 some candidates sought to air political advertising that showed aborted fetuses. They relied upon rules that limited station rights to edit political spots. *Broadcasting & Cable*'s Harry Jessell reported in

1994: "The FCC staff has proposed allowing television stations to relegate political spots containing graphic depictions of aborted fetuses to times when relatively few children are watching TV, commission sources say."[9]

> Following protest from stations, which feared the impact the ads would have on viewers and advertisers, the FCC informally ruled stations might deem the ads indecent and channel them to the 8 P.M.–6 A.M. "safe harbor" for indecent programming. Stations also were permitted to air warnings that the ads might be unsuitable for viewing by children.

The proposed rule would accomplish the same thing. But according to one source, stations would no longer have to decide whether the ads are indecent. They would be able to channel any ad they believe would be "harmful to children," the source said.[10]

In December 1994 *Broadcasting & Cable* magazine reported that the official ruling balanced the station's right to channel against a candidate's right to free political speech.

FCC Allows Movement of Antiabortion Ads

The long-awaited ruling on graphic antiabortion advertisements emerged from the FCC Nov. 22, two weeks after the national elections. Stations may "reschedule or channel political advertisements containing graphic abortion imagery to time periods when children are less likely to be in the audience." However, the FCC said, a station cannot move a spot simply because it does not like the ad's political message.

Source: *Broadcasting & Cable*, December 5, 1994, p. 62.

Fair access to the limited time and space of the broadcast medium was defined by the Supreme Court's ruling in *Red Lion Broadcasting Co. v. FCC*, 395 U.S. 367 (1969). However, the Court clarified in *CBS v. Democratic National Committee*, 412 U.S. 94 (1973) the right of broadcasters to exercise judgment—even where some free expression might be limited:

> Journalistic discretion would be in many ways lost to the rigid limitations that the First Amendment imposes on government. Application

of such standards to broadcast licensees would be antithetical to the very ideal of vigorous, challenging debate on issues of public interest. Every licensee is already held accountable for the totality of its performance of public interest obligations.[11]

Currently, broadcasters face a problem when a candidate seeks to use the medium for political speech that may be deemed "offensive." If they grant an unconditional right of access, then they run the risk of facing indecency complaints. If they regulate content, they run the risk of violating political rules.

Following the death of the Fairness Doctrine in the late 1980s, broadcast managers may have been confused. It could have seemed that all fairness obligations were removed by the FCC. In fact, however, the FCC did not have the authority to remove the fairness clause from political rules of Secs. 312 and 315 of the Communications Act.

Issues of balance might also surface for nonpolitical advertising. The rules of commercial speech law are less clear, but some generalizations are possible. Advertisers clearly have some rights to use the public airwaves, and they should have the expectation of being treated fairly by the broadcaster. In particular, the FCC has relied on the "reasonable judgment" of broadcasters in their decision making. The government has been most concerned about false, deceptive, or misleading advertising, and those spots promoting illegal activities or products.[12] In the case of indecency in advertising, the government would be forced to show it has a "substantial" interest—namely, protection of children.[13]

Advertising as Nonentertainment Programming

A classic view of advertising is expressed by David Ogilvy: "I do not regard advertising as entertainment or an art form, but as a medium of information."[14] Ogilvy takes a functional approach to the use of sex in advertising—only if it is relevant:

> Advertising reflects the mores of society, but does not influence them. Thus it is that you find more explicit sex in magazines and novels than in advertisements. The word fuck is commonplace in contemporary literature, but has yet to appear in advertisements.[15]

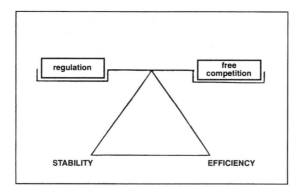

FIGURE 10.1   The economic marketplace vs. regulation—a study of the magic triangle. Source: Henri Louberge, *Risk, Information, and Insurance.* Courtesy of Kluwer Academic Publishers.

Ogilvy uses the example of a series of bikini advertisements in Paris in which a model promises and then removes her top and bottom— an event not shocking to the French. "But I would not advise you," Ogilvy warns, "to put up these posters in South Dakota."[16]

Notes

1. Sydney W. Head, Christopher H. Sterling, and Lemuel B. Scholfield, *Broadcasting in America*, 7th ed. Boston: Houghton Mifflin, 1994, pp. 229–263.
2. Ibid., p. 231.
3. Ibid., p. 259.
4. Robert G. Picard, *Media Economics: Concepts and Issues.* Newbury Park, Calif: Sage, 1989, p. 9.
5. Ibid., p. 26.
6. Ronald Berman, *Advertising and Social Change.* Beverly Hills: Sage, 1981, p. 52.
7. Ibid., p. 53.
8. Ibid., p. 55
9. Harry A. Jessell, "FCC moves on antiabortion ads; agency likely to permit 'channeling' of campaign spots depicting aborted fetuses; Federal Communications Commission, includes political broadcasting tips," *Broadcasting & Cable* 124(35), 29 August 1994, p. 46.
10. Ibid.
11. Donald M. Gillmor, Jerome A. Barron, Todd F. Simon, and Herbert A. Terry, *Mass Communication Law: Cases and Comment*, 5th ed. St. Paul: West Publishing, 1990, p. 514, quoting *CBS v. DNC*. Also see *Gillett Communications v. Becker*, 807 F.Supp. 757, 763 (N.D.Ga. 1992), where a lower court ruled that an Atlanta station could channel graphic abortion ads: "The Court is convinced that this time slot best accommodates the two competing interests and rights: the inter-

est in protecting children from indecent materials and Mr. Becker's right to broadcast his political advertisement." An earlier letter from the FCC had found that the ad was not indecent (See Letter, 7 FCC Rcd. No. 18, 5599-5560).

12. Don R. Pember, *Mass Media Law*, 6th ed. Dubuque, Iowa: WCB Brown & Benchmark, 1993, p. 503.

13. Ibid.

14. David Ogilvy, *Ogilvy on Advertising*. New York: Vintage Books, 1983, p. 7.

15. Ibid., p. 26.

16. Ibid.

Chapter 11

United States Court of Appeals for the District of Columbia Circuit Influence: The ACT Cases and Regulatory Ambiguity

Professional broadcasters may make the mistake in believing that the United States Supreme Court is where the law of broadcast indecency is found. In fact, the *Pacifica* decision, which failed to settle the matter, has left much to the United States Court of Appeals, District of Columbia Circuit, to deal with in the 1980s and 1990s.

The "politics" of broadcast regulation—namely, the influence of various political players—can be seen clearly in the case of broadcast indecency policy.[1] While the Federal Communications Commission (FCC), Congress, the White House, the United States Supreme Court, citizens groups, and industry lobbyists have played a part in the unfolding drama, the thesis of this chapter is that the United States Court of Appeals for the District of Columbia Circuit (D.C. Circuit) has played the most important role in the evolution of broadcast indecency policy. The purpose of this chapter, then, is to explore the rulings of the D.C. Circuit during the past decade.

Historical Context of Broadcast Regulation

Mass-media historian Louise Benjamin has found that even before the Radio Act of 1927,[2] American courts were influential in defining broadcast regulatory principles.[3] When a state court protected the signal of WGN, Chicago, from interference, the decision was "hailed as a means of clearing up the ether."[4]

Prior to 1927, "Congress had passed only two laws dealing with radio: the Wireless Ship Act of 1910[5] and the Radio Act of 1912.[6] However, because the laws were not aimed at mass audience broadcasting, Secretary of Commerce Herbert Hoover found that self-regulation of frequency usage was not working.[7] But before the Second Radio Conference of 1923,[8] the courts stepped in stating: "Shortly before the conference, Hoover's attempts to regulate were seriously undermined when the U.S. Court of Appeals for the District of Columbia Circuit ruled that the secretary of commerce lacked legal discretion to withhold licenses from broadcast stations."[9]

It is clear, therefore, that before there ever was a Federal Radio Commission or an FCC, the political struggle over who would control the airwaves, as well as how they would be regulated, was underway. It can be said that many of the defining moments in broadcast regulation policy came through judicial decision making rather than from Congress or the FCC. In *National Broadcasting Company v. United States*,[10] for example, the United States Supreme Court dealt with the "public interest" standard[11] and the right of the FCC to manage the public airwaves.[12] Later, in *Red Lion Broadcasting Company v. FCC*, it was argued that the rights of viewers and listeners were more important than the rights of broadcasters.[13] In the area of broadcast indecency, a sharply divided Court in *FCC v. Pacifica Foundation* upheld FCC attempts to consider the content of broadcast speech on a case-by-case basis.[14] But lack of clarity in that opinion, coupled with FCC ambiguity, led to a twenty year period of regulatory confusion. In such a climate, it should not be surprising that the courts, and specifically the D.C. Circuit, became major participants in the political process.

Politics and the D.C. Circuit

The influence of the D.C. Circuit appeared to become increasingly important in the late 1960s with respect to broadcast regulatory policy.[15] First Amendment challenges—especially those core challenges that the United States Supreme Court has tended to avoid—have often been considered by three-judge panels of the D.C. Circuit.[16] Commentators such as Erwin G. Krasnow, Lawrence D. Longley, and Herbert A. Terry argued that "the vague public interest standard embodied in the Communications Act by Congress has offered the courts the opportunity for a significant role in overseeing the FCC."[17]

The D.C. Circuit and Broadcast Indecency Policy

In order to understand the influence of the D.C. Circuit on broadcast indecency policy, the court's reaction to other political institutions in a series of cases beginning in the late 1980s must be examined.

Act I

The D.C. Circuit attempted to sort out the post-*Pacifica* world of broadcast indecency in *Action for Children's Television v. FCC (Act I)*.[18] The 1988 *Act I* case held that the FCC had changed its enforcement standard in 1987, stating, "We uphold the generic definition the FCC has determined to apply, case-by-case, in judging indecency complaints, but we conclude that the Commission has not adequately justified its new, more restrictive channeling approach, *i.e.*, its curtailment of the hours when non-obscene programs containing indecent speech may be broadcast."[19]

In an opinion filed by Circuit Judge Ruth Bader Ginsburg, the court seemed to remind the FCC that indecent speech is protected by the First Amendment, and the "avowed objective is not to establish itself as censor but to *assist parents* in controlling the material young children will hear."[20] Although the D.C. Circuit appeared constrained by the precedent of *Pacifica* with respect to vagueness challenges to the indecency policy, the court volleyed the issue back to the FCC by holding that there was insufficient evidence to support time channeling to late-night hours as an effective method to protect children.[21] The court found that while indecency has First Amendment protection, the FCC may regulate children's access to it.[22] Specifically, the court stated: "Broadcasting is a unique medium; it is not possible simply to segregate material inappropriate for children, as one may do, *e.g.* in an adults-only section of a bookstore. Therefore, channeling must be especially sensitive to the First Amendment interests of broadcasters, adults, and parents."[23]

The D.C. Circuit utilized the decision as a vehicle to tell the FCC it "would be acting with utmost fidelity to the First Amendment were it to reexamine, and invite comment on, its daytime, as well as evening, channeling prescriptions."[24] The court instructed the FCC that it needed evidence to support a rule for promoting parental, not government control: "A securely grounded channeling rule would give effect to the government's interest in promoting parental supervision of children's listening, without intruding excessively upon the licensee's range of discretion or the fare available for mature audiences and even children whose parents do not wish them sheltered from indecent speech."[25] The *Act I* case, however, settled little and was only the beginning of the D.C. Circuit's attempt to influence the political process.

Act II

In the 1991 case of *Action for Children's Television v. FCC (Act II)*, the D.C. Circuit upheld its *Act I* decision in spite of an FCC twenty-four-hour ban ordered by Congress.[26] The court said it had ordered the FCC to hold hearings and determine when stations could broadcast indecency, but "[b]efore the Commission could carry out this court's mandate, Congress intervened."[27] Two months after the *Act I* decision, the 1989 funding bill contained a "rider" requiring the FCC to enforce indecency regulation *"on a 24 hour per day basis."*[28] Faced with new orders, the FCC abandoned plans to follow the *Act I* orders.[29] Then, in 1989, the United States Supreme Court in *Sable Communications, Inc. v. FCC* rejected a "blanket ban on indecent commercial telephone message services" at the same time as distinguishing dial-a-porn services from broadcasting.[30]

The *Act II* court restated its *Act I* admonishment, declaring, "Broadcast material that is indecent but not obscene is protected by the first amendment; accordingly, the FCC may regulate such material only with due respect for the high value our Constitution places on freedom and choice in what the people say and hear."[31] The court additionally stated that "[w]hile 'we do not ignore' Congress' apparent belief that a total ban on broadcast indecency is constitutional, it is ultimately the judiciary's task, particularly in the First Amendment context, to decide whether Congress has violated the Constitution."[32]

The court rationalized that Congress's action came before the *Act I* decision, "thus, the relevant congressional debate occurred without the benefit of our constitutional holding in the case."[33] The court argued that the precedent of *Act I* and of *Sable Communications* guaranteed adult access to indecency and limited regulation to that which would "restrict children's access."[34] The court agreed with one FCC commissioner who had called the mandate unconstitutional.[35] The court stated, "[N]either the Commission's action prohibiting the broadcast of indecent material, nor the congressional mandate that prompted it, can pass constitutional muster under the law of this circuit."[36] Then the court spoke directly to the political tangle the FCC found itself in over the blanket ban:

> We appreciate the Commission's constraints in responding to the appropriations rider. It would be unseemly for a regulatory agency to throw down the gauntlet, even a gauntlet grounded on the Constitution, to Congress. But just as the FCC may not ignore the dictates of the legislative branch, neither may the judiciary ignore its independent duty to check the constitutional excesses of Congress. We hold that Congress' action here cannot preclude the Commission from creating a safe harbor exception to its regulation of indecent broadcasts.[37]

The court had flexed its political muscle and cloaked it in judicial responsibility. The D.C. Circuit clarified that, even though it was the Congress that had original responsibility of regulating broadcasting as interstate commerce, and had delegated that authority to the FCC, it was the D.C. Circuit which was charged with protecting the First Amendment of the United States Constitution.[38] While the court was by no means staking out absolutist ground, it was bending over backward to fashion a limited regulatory scheme—one that still would need to be supported by forthcoming data. In the end, the remand of the *Act II* case volleyed the political ball back into the court of the FCC and set the stage for *Act III*.

Act III

A three-judge panel of the D.C. Circuit again reviewed broadcast indecency regulation in 1993, in *Action for Children's Television v.*

FCC (Act III). In *Act III,* a group of broadcasters, programmers, listeners, and viewers had challenged a provision in the Public Telecommunications Act of 1992—the public broadcasting funding bill—which directed the FCC to ban indecent material between 6 A.M. and midnight.[39]

In *Act III,* the court refused to accept the notion that much had changed since its previous decisions, stating, "While we break some new ground, our decision that the ban violates the First Amendment relies principally upon two prior decisions of this court in which we addressed similar challenges to FCC orders restricting the broadcasting of 'indecent' material, as defined by the FCC."[40] In reviewing the FCC's 1993 implementation order, the D.C. Circuit agreed that children need to be protected from indecency and that parents might need help from the government in protecting their children from indecent broadcasts.[41] However, the D.C. Circuit rejected the idea—restated from *Pacifica*—that children *and* adults need to be protected from "indecent material in the privacy of their homes."[42] The court stated, "we accept as compelling the first two interests involving the welfare of children, but in our view, the FCC and Congress have failed to tailor their efforts to advance these interests in a sufficiently narrow way to meet constitutional standards."[43] The D.C. Circuit then identified its curious political position as a buffer between FCC actions and Supreme Court interpretations: "While *Act I* acknowledges that *Pacifica* 'identified' an interest in 'protecting the adult listener from intrusion, in the form of offensive broadcast materials, into the privacy of the home,' it does not endorse its legitimacy."[44]

The *Act III* court, rather than emphasizing the narrow First Amendment view of *Pacifica,* took a much broader position. The government has no general interest, the court wrote, in protecting *adults* "primarily because the official suppression of constitutionally protected speech runs counter to the fundamental principle of the First Amendment 'that debate on public issues should be uninhibited, robust, and wide-open.' "[45] It is significant that the court, on this point, chose to select a print media case to interpret the First Amendment. The suggestion is made in the first three *Act* cases that a narrow regulatory slice has been carved—one that will only be justified when the governmental interest of protecting *children* is supported with hard data. The burden is on the government, and it is substantial. Even if the case can be made, the opinion accepts the notion of parental responsibility:

Viewers and listeners retain the option of using program guides to select with care the programs they wish to view or hear. Occasional exposure to offensive material in scheduled programming is of roughly the same order that confronts the reader browsing in a bookstore. And as a last resort, unlike residential picketing or public transportation advertising, "the radio [and television] can be turned off."[46]

The D.C. Circuit struck a solid blow to the foundation of broadcast regulation in its view that broadcast speech has core First Amendment value. In challenging the notion of intrusion of broadcast signals into one's home, the court pointed out that listeners and viewers have controls that can be exercised without turning to the government.[47] Left with the government interest in protecting children from broadcast indecency, the D.C. Circuit restrained the FCC by applying a "least restrictive means" test.[48] Any ban would have to survive such a test, and according to the court, "the government did not properly weigh viewers' and listeners' First Amendment rights when balancing the competing interests in determining the widest safe harbor period consistent with the protection of children."[49] While some sort of safe harbor might survive judicial scrutiny, the court wrote: "[W]e are at a loss to detect any reasoned analysis supporting the particular safe harbor mandated by Congress."[50] As a matter of political power, the court of appeals effectively stopped Congress and its administrative agency in their regulatory tracks.

On the issue of parental supervision and the validity of a safe harbor, the court clearly rejected the FCC argument when it wrote: "[O]ne could intuitively assume that as the evening hours wear on, parents would be better situated to keep track of their children's viewing and listening habits."[51] The FCC argument is grounded in the notion that parents cannot effectively supervise the television and/or radio habits of their children. The court's response is clear:

> [T]he government has not adduced any evidence suggesting that the effectiveness of parental supervision varies by time of day or night, or that the particular safe harbor from midnight to 6 A.M. was crafted to assist parents at specific times when they especially require the government's help to supervise their children. The inevitable logic of the government's line of argument is that indecent material can never be broadcast, or, at most, can be broadcast during times when children are surely asleep; it could as well support a limited 3:00 A.M. to 3:30 A.M. safe harbor as one from midnight to 6 A.M.[52]

The protection of children argument is further tempered by the conclusion that *Pacifica* addressed only the need to protect children under the age of twelve.[53] The FCC, instead, had attempted to treat "teens aged 12–17 to be the relevant age group for channeling purposes" in the *Act* cases.[54] But the D.C. Circuit noted: "When the government affirmatively acts to suppress constitutionally protected material in order to protect teenagers as well as younger children, it must remain sensitive to the expanding First Amendment interests of maturing minors."[55] Circuit Judge Harry Edwards, in a concurring opinion, toyed with the complicated issue of indecency regulation. Beyond not knowing what effects indecent content might have on which children, Edwards considered what would happen if "most parents would prefer to retain the right to decide."[56]

> [C]ould Congress still ban the showing of indecent material? If so, on what terms? Would it be prompted by a "moral judgment" that indecent material is bad for all children of all ages? And, if so, how can that be squared with the Supreme Court's rulings that distinguish between unprotected "obscene" and protected "indecent" materials, and suggest that the ages of minors must be considered in assessing the vulnerability of children?[57]

Judge Edwards argued that the government's interest "is tied directly to the magnitude of the harms sought to be prevented."[58] Yet the FCC failed to show "precisely what those harms are."[59] In short, the *Act III* opinion might have been a powerful weapon against *any* broadcast indecency regulation—had the panel's 1993 decision not been vacated in 1994. Instead, as we will see, a regulatory position re-emerged.

Interpretations of the Opinion

Legal scholar Jeffrey Stein isolated four questions emerging from the *Act III* decision:

- Is the "generic" definition of indecency too vague?
- Should the FCC pursue establishing "safe harbors" for indecent content?
- Should the FCC pursue a total ban on broadcasting indecency?
- Would the FCC be better served by pursuing case-by-case enforcement of the generic indecency definition instead?[60]

Stein argued that vagueness should continue to be challenged by broadcasters, that a safe harbor is without empirical support, that a total ban is not constitutional, and that case-by-case decisions run the risk of being found inconsistent.[61] Surprisingly, Stein's proposal called for abandonment of content regulation in favor of a return to "criminal prosecution . . . rather than administrative agency proceedings."[62] He wrote: "[I]t would allow triers of fact to review local standards in local communities to determine if violations have occurred."[63] However, proposing that criminal prosecutions are a solution to the indecency quagmire seems to be an admission that the current policy system is a mess; and suggesting that the FCC should retreat to license renewal analysis in such indecency violations, models a weaker role for the FCC. Stein's view fails to recognize that the FCC may serve an important function in *protecting* broadcasters from indecency complaints.

One cannot assume that the FCC, through its tangled regulatory process, has failed, in the end, to protect the First Amendment rights of free speech for broadcasters. The regulation of broadcast indecency occurs, not in local communities of interest, but in our nation's capital. Locked in the political milieu that is Washington, D.C., an offending broadcaster may escape direct scrutiny. Even where FCC review leads to a fine, these economic sanctions rarely can be seen as significant to corporate group owners. When Infinity Broadcasting President and CEO Mel Karmazin agreed in late 1995 to pay $1.7 million to settle Howard Stern's indecency complaints—a move designed to clear the record for a new round of multimillion-dollar transactions—he told *Broadcasting & Cable* magazine: "we want to have a good relationship with the government without in any way, shape, or form compromising what we believe to be our First Amendment rights" (September 11, 1995, at 9).

Political Generalizations

The authors of *The Politics of Broadcast Regulation* identified seven generalizations about regulatory policy making.[64] These may help us to analyze recent developments and make predictions about future action. The case of shock jock Howard Stern is a recent example of the process.

1.   *Participants seek conflicting goals from the process.* In the case of broadcast indecency regulation, not everybody can be a winner. The protection of children, if possible, would come at the expense of diminishing free speech rights for broadcasters and adult listeners. The various positions ranging from absolute free speech to a total ban suggest the political reality that compromise with the broadcast industry is likely. The FCC and Infinity can settle their public dispute while First Amendment loyalists continue obscure legal battles in the federal courts.

2.   *Participants have limited resources insufficient to continually dominate the policy-making process.* Broadcasters interested in challenging FCC regulatory initiatives must make an economic decision about the value of their actions. Likewise, programmers must weigh their options. The sheer slow pace of regulatory change is in stark contrast to rapid media change. For example, the FCC Infinity settlement, according to Karmazin, will lead to new business opportunities: "Now we feel there will be many, many more broadcasters interested in taking Howard's show into many more markets than he has been in up to now."

3.   *Participants have unequal strengths in the struggle for control or influence.* The court of appeals, largely because the Supreme Court has avoided further significant review of broadcast indecency, holds the position of "court of last resort." However, the court of appeals' authority ends with the publication of its decisions. During the current decade-long struggle, the FCC has refused the court's suggestion to collect and analyze hard data on damaging effects. The FCC, to its credit, recognized that media effects research results have been inconclusive. The latest round of decision making in mid-1995, as we will see, appears to acknowledge that the FCC is the administrative agency that must, in the end, answer to Congress on broadcast indecency.

4.   *The component subgroups of participant groups do not automatically agree on policy options.* The absolutist First Amendment view of Howard Stern's broadcast group, as well as others representing shock-jock deejays, is not shared by all broadcasters. In fact, there have been those who have argued that such blue radio is bad for the long-term health of the industry. Likewise, members of the court of appeals and the FCC have disagreed over the years about free speech rights. The *Pacifica* decision of the Supreme Court is perhaps the best example of division. Infinity Broadcasting continues to hold the position in court that FCC indecency rules are unconstitutional.

5. *The process tends toward policy progression by small or incremental steps rather than massive changes.* In a sense, the dispute over broadcast indecency arose because the FCC attempted something larger than incremental policy change in the late 1980s. The reaction from interest groups was swift. The judicial review slammed the brakes on any attempt at massive change in policy. In one round of decision making, the question on the table was simply whether a "safe harbor" should begin at 8 P.M., 10 P.M., or midnight.

6. *Legal and ideological symbols play a significant role in the process.* The symbol of children as being defenseless against indecent broadcast is perhaps the most potent one in this process. Freedom and autonomy are also important ideological symbols in the indecency debate. Precedent is perhaps the most significant legal symbol, and it surfaces when the court of appeals expresses being bound by it. Likewise, judicial review is an important legal symbol in the process.

7. *The process is usually characterized by mutual accommodation among participants.* Early on, it was difficult to see much mutual accommodation on broadcast indecency. As a highly polarized issue, the middle ground for compromise seemed difficult to discover. But developments in 1995 did, as the political model predicts, lead participants toward accommodation. In the case of Howard Stern's broadcasts, Infinity won a clean, expunged record, and the FCC won the public perception that they were protecting children by regulating the public airwaves.

Recent Developments

By May 1995, the FCC was poised to clarify its current policy on broadcast indecency. *Broadcasting & Cable* magazine wrote:

> The FCC is putting the finishing touches on a broadcast indecency report it hopes will give TV and radio stations a better idea of what's actionable, agency officials say. The FCC agreed to write the report as part of last year's settlement of an indecency case against Evergreen's WLUP(AM) Chicago. Since Chairman Reed Hundt

arrived in November 1993, the FCC has not aggressively enforced the indecency statute. But it hasn't ignored it, either. Indeed, late last Friday it slapped WGRF (AM) Buffalo with a $4,000 fine for an off-color 1993 broadcast.[65]

It has long been an issue of whether there is a legitimate dichotomy between what the FCC judges as "actionable" and "nonactionable" with respect to pending complaints. To this point, the FCC's position has been that broadcasters should be able to see a distinction, but the case-by-case review policy tended to increase ambiguity. Further complicating matters is the fact that broadcast indecency policy revision comes as the United States Congress debates a series of controversial broadcast dereg-ulation proposals.[66]

Political Implications of Recent Court Actions

Without significant action from the United States Supreme Court on broadcast indecency policy, it appears that the D.C. Circuit will continue to hold an upper hand in setting long-term bound-aries for free broadcast speech. In a 7-4 decision in the summer of 1995 (*Act IIIb*), the Court of Appeals granted the FCC authority to channel indecent broadcasts from 10 P.M. to 6 A.M. local time (*Broadcasting & Cable*, July 3, 1995, at 10). "Parents and the public are the winners," FCC Chair Reed Hundt told the press. First Amendment lawyers said they were "deeply disappointed." The Court of Appeals wrote, "We are dealing with questions of judg-ment; and here, we defer to Congress's determination of where to draw the line. . . ." (Id.) A *Broadcasting & Cable* editorial said the decision was a reminder that broadcasters are "second-class citi-zens in terms of the First Amendment" and called for Supreme Court review (Id., at 50). The Court of Appeals followed the deci-sion with yet another ruling (*Act IV*) in July 1995 that upheld lengthy FCC review of complaints—from nine months to seven years (*Broadcasting & Cable*, July 24, 1995, at 65). Hundt said the ruling "further empowers parents to shield children from inde-cent programming" (Id.). It was clear that the support for FCC regulation by the Court of Appeals probably was a factor leading to the Infinity settlements (*Broadcasting & Cable*, September 4, 1995, at 6). Without insulation from the Court of Appeals, broadcasters will face FCC regulation driven by a climate of political pressures.

Politics of Broadcast Regulation in the 1990s

The generalizations made in *The Politics of Broadcast Regulation* seem to apply well to the case of broadcast indecency policy-making in the 1990s. Still, one can argue that a systems model approach for understanding the process favors description over prediction. What is needed is more comprehensive theory building in the area of normative media concerns. Any political model would need to build upon social theory that would help predict how regulation functions on an economic landscape.

Implications for Future Study of Policy Making

Future research on broadcast indecency regulations should recognize previous generalizations and begin to link them to larger social theories of mass communication. Missing from most previous analyses is a grounding in social theory. The emphasis has been on summarizing and describing court decisions. These legal analyses fall short of providing an understanding of the law in a social context.[67] Legal commentators would do well to look to law reviews and scholarly communication journals for analyses that link broadcast indecency regulation to what we know about governmental and social control of communication messages. In addition, much has been made of deregulation through technological innovation. For example, Edwin Diamond, Norman Sandler, and Milton Mueller, long before the V-chip (*Broadcasting & Cable*, June 26, 1995, at 16), argued that scrambling devices could be employed to protect children from harmful media messages.[68]

If reason is to be brought to bear on broadcast indecency policy, then "deregulation" must be distinguished from "policy making." In the words of one analyst: "[C]ommunications deregulation lacks not only an agreed upon definition, but also an agreed-upon goal."[69] The future of deregulation and policy making should be grounded in historical First Amendment free speech principles and theoretic predictions about the limitation of content regulation in a free society.

Manager's Summary

Broadcast managers continue to face a murky political process with respect to broadcast indecency. They must rely upon Washington lawyers to interpret policy. Future deregulation should move forward with free speech principles in mind.

Notes

1. Erwin G. Krasnow, et al., *The Politics of Broadcast Regulation*, 3rd ed. New York: St. Martin's Press, 1982.
2. 47 U.S.C. § 81 et seq. (repealed 1934). Section 109 of the Act states in pertinent part that "no person within the jurisdiction of the United States shall utter any obscene, indecent, or profane language by means of radio communication." 47 U.S.C. § 109, quoted in Milagros Rivera-Sanchez, "Developing an Indecency Standard, The Federal Communications Commission, and the Regulation of Offensive Speech, 1927–1964," 20(1) *Journalism History* 3 (1994). The *Radio Act of 1927* was repealed in 1934, but this prohibition was transferred into the *Communications Act of 1934*. Milagros, at 3–4. Later, in 1948 this same clause was incorporated into § 1464 of Title 18 of the United States Code. Id. at 4. Section 1464 attached a punishment to prohibition of a "fine of not more than $10,000 or imprisonment of not more than two years or both." Id. (quoting 18 U.S.C. 1464 (1984)).
3. Louise M. Benjamin, "The Precedent that Almost Was: A 1926 Court Effort to Regulate Radio," 67(3) *Journalism Quarterly* 578–585 (1990). The Benjamin article discusses the impact of a case decided prior to the enactment of the Radio Act of 1927.
4. Ibid. at 583. The American Bar Association, the National Radio Coordinating Committee and newspapers reacted positively to the decision.
5. Milagros Rivera-Sanchez, "The Origins of the Ban on 'Obscene, Indecent, or Profane' Language of the Radio Act of 1927" in 149 *Journalism Monographs* 1 (1995). The *Wireless Ship Act* failed to address the issue of offensive language used by amateur operators.
6. Krasnow supra note 1, at 10. See also, Milagros Rivera-Sanchez, "Developing an Indecency Standard, The Federal Communications Commission, and the Regulation of Offensive Speech, 1927–1964," 20(1) *Journalism History* 3–14 (1994) (tracing history of FCC's regulation of broadcast prior to *Pacifica*).
7. Krasnow supra note 1, at 11.
8. Hoover called the Second Radio Conference in 1923 to address the problem of reception interference caused by the crowding of stations.
9. Ibid. at 11.
10. 319 U.S. 190 (1943).
11. Ibid. at 216. The "public interest" standard was set forth in the *Communications Act of 1934* as a criterion for the exercise of power by the FCC. The Court stated that the determination of whether a particular regulation served the public inter-

est was to be interpreted "by its context, by the nature of radio transmission and reception, by the scope, character, and quality of services . . ." Ibid. (quoting *Federal Radio Commission v. Nelson Bros. Bond & Mortgage Co.*, 289 U.S. 266 (1933).

12. Prior FCC action focused on "technical and engineering impediments to the 'larger and more effective use of the radio in the public interest.' " Ibid. at 217. The Court found, however, that nothing in the *Communications Act* precluded the FCC from exercising licensing and regulatory powers consistent with the "public interest" standard.

13. 395 U.S. 367 (1969). The Court held that an FCC promulgated "fairness doctrine" requiring broadcasters to afford political candidates, who have been criticized over the broadcaster's facilities, an equal opportunity to respond, did not violate the First Amendment. The Court reasoned that "[i]t is the right of the viewers and listeners, not the right of the broadcasters which is paramount." Ibid. at 390.

14. 438 U.S. 726 (1978). At issue in *Pacifica* was the FCC's determination that a certain broadcast aired during the afternoon was "patently offensive." The FCC issued a declaratory warning order, stating that future broadcasts of the type at issue could result in the assessment of penalties. The majority defined indecency as a "function of context," not to be "judged in the abstract." Ibid. at 742. Consequently, the FCC's decision to characterize the broadcast as "patently offensive" rested "on a nuisance rationale under which context is all important." Ibid. at 750.

15. Krasnow, supra note 1 at 63. This increasing importance is a result of citizen groups raising questions "that have never been subjected to the crucible of judicial review."

16. Ibid.

17. Ibid. at 64. The courts have been given the responsibility of determining whether the FCC has properly acted within the public interest standard.

18. 852 F.2d 1332 (D.C. Cir. 1988) [hereinafter *Act I*]. In December 1987, the FCC issued a Reconsideration Order, which affirmed three FCC rulings and restricted the safe-harbor period to exclude the hours between 10:00 P.M. and 12:00 midnight. Ibid. at 367. Previously, the safe-harbor period was from 12:00 midnight until 6:00 A.M. Id. The three rulings at issue concerned previous FCC determinations that certain aired material was indecent, even though the material did not violate the test developed in *Pacifica*. See Pacifica Foundation, 2 F.C.C.R. 2698 (1987); Regents of the University of California, 2 F.C.C.R. 2703 (1987); Infinity Broadcasting Corp., 2 F.C.C.R. 2705 (1987).

19. *Act I*, 852 F.2d 1332, 1334 (D.C. Cir. 1988). Petitioners, consisting of commercial broadcasting networks, public broadcasting entities, licensed broadcasters, associations of broadcasters and journalists, program suppliers, and public interest groups, challenged the constitutionality of the order. Id. Petitioners argued that the new safe harbor period violated the First Amendment by denying adults access to constitutionally protected material. Specifically, petitioners alleged that the order was facially invalid on the grounds of vagueness and overbreadth and that the order was arbitrary and capricious. Id. The court vacated and remanded those cases involving post-10:00 P.M. broadcasts, finding that the FCC failed to offer sufficient evidence in support of the new safe-harbor period.

20. Ibid. (emphasis added). The FCC argued that the government's interest was limited to "protecting unsupervised children from exposure to indecent material." Id. at 1343. The court voiced its concern by instructing the FCC to formulate a channeling rule to promote parental control rather than government censorship.

21. Petitioners argued that the new standard was " 'inherently vague' and was installed without any evidence of a problem justifying a thickened regulatory response." Ibid. at 1338 (citing Brief of Petitioners at 39).

22. Specifically, the court held that the "power of the state to control the conduct of children reaches beyond the scope of its authority over adults . . ." Ibid. at 1340 (quoting *Ginsburg v. New York*, 390 U.S. 629, 638 (1968)). Consequently, the court reasoned that the offensive impact of certain words or phrases on children will not necessarily be overridden by the overall social value of the program.

23. *Act I*, 852 F.2d at 1340 n.12. Historically, regulation of the broadcast industry has been based on the unique characteristics of the medium such as scarcity of resources. See *Red Lion v. FCC*, 395 U.S. 367 (1969). The broadcast medium is subject to a scarcity of resources. Ibid. at 396. Due to a limited number of available frequencies, not all those speakers who wish to gain access to broadcast have the capability. Id. Thus, the FCC has utilized the "scarcity rationale" to justify its regulation of the broadcast industry.

24. *Act I*, 852 F.2d at 1341.

25. Ibid. at 1344.

26. 932 F.2d 1504 (D.C. Cir. 1991), cert. denied, 503 U.S. 913 (1992) [hereinafter *Act II*]. Two months following the *Act I* decision, an appropriations bill, requiring the FCC to promulgate regulations to enforce the ban on indecent material on a 24 hour a day basis was signed into law. Id. at 1507. See also Pub. L. No. 100–259, § 608 (1984).

27. *Act II*, 932 F.2d 1504, 1507. In following Congress's directive, the FCC promulgated the new rule banning all broadcasts of indecent material and abandoned its plans to investigate the issues raised by the court in *Act I*.

28. Citing Pub. L. No. 100-459, § 608, 102 Stat. 2228 (1988) (emphasis added).

29. The court recognized that the FCC is bound by the orders of the legislative branch. Ibid. at 1509.

30. *Sable Communications Inc. v. FCC*, 492 U.S. 115 (1989) (total ban on indecent telephone communications did not pass strict scrutiny because not narrowly tailored).

31. *Act II*, 932 F.2d at 1508 (citing Act I, 852 F.2d at 1344). For a complete discussion of *Act I*, see supra notes 18–25 and accompanying text.

32. Ibid. at 1509.

33. Thus, despite the congressional mandate banning broadcast indecency, the court asserted that such a prohibition cannot survive constitutional scrutiny.

34. The court looked to the Supreme Court's decision in *Sable Communications* in support of its affirmation that indecent material does receive First Amendment protection and a total ban on indecent material does not comport with the Constitution.

35. Prior to Congress's enactment of the appropriations rider, the FCC was of the view that if 1464 were to be read as imposing a total ban on indecent broadcast, it would be a constitutional violation of the FCC's authority.

36. *Act II*, 932 F.2d 1504. In order for a regulation to be constitutionally permissible, it must be carefully tailored to serve a compelling government interest. Id. at 1509. The government carries the burden of establishing a compelling interest.
37. Ibid. at 1509–1510. This decision returned the FCC to the same position it was in following the *Act I* decision. Id. The court concluded its opinion by instructing the FCC to resume its plans to address the concerns raised by the court in *Act I*. Id. See also *Act I*, 852 F.2d 1332 (detailing discussion of concerns to be addressed by FCC).
38. *Act II*, 932 F.2d at 1509.
39. *Action for Children's Television v. FCC*, 11 F.3d 170 n.1 (D.C.Cir. 1993), vacated in *Action for Children's Television v. FCC*, 15 F.3d 186 (D.C.Cir. 1994) (responding to the Public Telecommunications Act of 1992, Pub.L. No.102–356 16(a), 106 Stat. 949, 954, (codified as *In Re Enforcement of Prohibitions Against Broadcast Indecency* in 18 U.S.C. 1464, 8 F.C.C.R. 704 (1993) (1993 Order))). Specifically the provision reads:
   BROADCASTING OF INDECENT PROGRAMMING
      SEC. 16. (a) FCC REGULATIONS—The Federal Communications Commission shall promulgate regulations to prohibit the broadcasting of indecent programming—
      (1) between 6 A.M. and 10 P.M. on any day by any public radio station or public television station that goes off the air at or before 12 midnight; and
      (2) between 6 A.M. and 12 midnight on any day for any radio or television broadcasting station not described in paragraph (1).
      The regulations required under this subsection shall be promulgated in accordance with section 553 of title 5, United States Code, and shall become final not later than 180 days after the date of enactment of this Act.
   106 Stat. at 954.
40. *Act III*, 11 F.3d at 171. For a discussion of the prior decisions see supra notes 18–37 and accompanying text.
41. Ibid. at 177.
42. Ibid. at 171. The FCC asserted three goals to justify its 1993 Implementation Order: (i)"ensuring that parents have an opportunity to supervise their children's listening and viewing of over-the-air broadcasts," (ii)"ensuring the well being of minors regardless of parental supervision," and (iii) protecting "the right of all members of the public to be free of indecent material in the privacy of their own homes." Id. (quoting *In Re Enforcement of Prohibitions Against Broadcast Indecency* in 18 U.S.C. § 1464, 8 F.C.C.R. 704, 705–706, ¶¶ 10, 14 (1993)).
43. Ibid. at 171.
44. Ibid. at 174 n.6 (citing *Act I*, 852 F.2d at 1344 n. 20).
45. *Act III*, 11 F.3d at 175, (citing *New York Times v. Sullivan*, 376 U.S. 254, 270 (1964)). The court went on to note that the First Amendment protects the rights of all listeners and viewers, "not just of that part of the audience whose listening and viewing habits meet with governmental approval." Thus, as long as obscenity is not involved, "[the Supreme Court] ha[s] consistently held that the fact that protected speech may be offensive to some does not justify its suppression." (citing *Bolger v. Youngs Drug Prods. Corp.*, 463 U.S. 60, 71 (1983)).

46. Ibid. at 176 (citing *Packer Corp. v. Utah*, 285 U.S. 105, 110 (1932)).
47. Ibid.
48. May want to state the "least restrictive means" test.
49. *Act III*, 11 F.3d at 177. See also Id. n.12. ("As conceded at oral argument, petitioners do not challenge the FCC's authority to regulate 'indecent' material in the broadcast media by creating a safe harbor outside of which indecent material may not be broadcast.")
50. Ibid. at 177.
51. Ibid. at 178. The FCC argued that "parents can effectively supervise their children only by co-viewing or co-listening, or, at a minimum, by remaining actively aware of what their children are watching and listening at all times." Id. (citing Respondents' Brief at 30–31). Because this type of supervision is not practical, the FCC also stated that "parents who seek to avoid exposing their children to indecency face a nearly impossible task if such material is broadcast" (citing Respondents' Brief at 26). However, the court noted that the FCC argument assumes that "regardless of the time of day or night, parents cannot effectively supervise their children's television or radio habits."
52. *Act III*, 11 F.3d at 178.
53. See Ibid. n.14. The FCC contended that *Pacifica* provided the government with the power to restrict the broadcast of indecent speech to minors regardless of age. Therefore, the FCC proposed that the First Amendment rights of minors need not be considered with regard to regulating indecent material in the broadcast media. Id. However, neither *Act I* nor *Pacifica* involved the First Amendment rights of teenagers. In addition, while the Supreme Court has stated that "the protection of children includes 'preventing minors from being exposed to indecent' material, it ultimately struck down the ban on indecent telephone messages because it was not narrowly tailored to serve that interest" (citing *Sable Communications Inc. v. FCC*, 492 U.S. 115, 131 (1989)).
54. Ibid.
55. Ibid. at 180. The court stated that at some point, "the government's independent interest in shielding children from offensive material no longer outweighs the First Amendment interests of minors in receiving important information" (citing *Erznoznik v. City of Jacksonville*, 422 U.S. 205, 213 (1975); *Bolger v. Youngs Drug Prods. Corp.*, 463 U.S. 60, 74–75 n. 30 (1983); *Carey v. Population Servs. Int'l*, 431 U.S. 678, 688–89 (1977)).
56. *Act III*, 11 F.3d at 185 (Edwards, J., concurring).
57. Ibid.
58. Ibid.
59. Ibid. In the FCC's 1993 Implementation Order, the FCC asserts that "harm to children from exposure to [indecent] material may be presumed as a matter of law." 1993 Order, 8 F.C.C.R. at 706–07, ¶¶ 17–18. The FCC also referred to studies demonstrating certain undefined "negative effects of television on young viewers' sexual development and behavior." However, the court stated that this evidence does not provide a significant basis for analyzing possible First Amendment intrusions. *Act III*, 11 F.3d at 185.
60. Jeffrey L.L. Stein, "The FCC's 'Indecent Proposal': A Drama in Three 'Acts,' " Association for Education in Journalism and Mass Communication, Law Div., annual conference, Atlanta (1994).

61. Ibid. at 36–45.
62. Ibid. at 46.
63. Ibid.
64. Krasnow, supra note 1, at 138–141.
65. "Indecency Guide," *Broadcasting & Cable*, 8 May 1995, at 113.
66. "Telcom Bill Advances in House," *Broadcasting & Cable*, 22 May 1995, at 10–11
67. See, e.g., Neal J. Friedman and Robert D. Richards, *Communications Law: Regulation of the Electronic Mass Media*, 340–341 (1995) (abstracting *Act III* but providing no commentary except a quote from Commissioner Quello suggesting FCC will ensure that Howard Stern and Infinity follow the law).
68. Edwin Diamond, et al., *Telecommunications in Crisis: The First Amendment, Technology and Deregulation*, Cato Institute, Washington, DC, (1983).
69. Jeremy Tunstall, *Communications Deregulation, The Unleashing of America's Communications Industry*, Oxford, UK: New York, (1986), p. 283.

Chapter 12

Broadcast Indecency and First Amendment Theory: The Future of Regulation in an International Context

To this point, our examination of the broadcast indecency issue has been centered on American law as developed through Congress, implemented by the FCC, and interpreted by the courts. However, in an increasingly global media environment, it is safe to say that the regulation of broadcast indecency will not be settled by any one nation. In an age of cross-border and international/global media, notions of decency will need to be reconsidered. We need to contrast the narrow regulatory issues of broadcast indecency against the broad notions of free speech.

Global Communication Freedom

One view of the future is that global communication will lead to greater freedoms:

> In time, given the global movement toward democracy, interactive voice, audio, video, and data exchanges will occur worldwide. The effects of such a market for goods, services, ideas, and information will be extraordinary. It will help break down barriers to free trade and market economies, will promote the transformation of totalitarian governments into democratic ones, and will

cause certain local cultures to become part of the international cultural melange.[1]

However, it is recognized that there are dangers: "If we are to avoid the losses in freedom that would be associated with content and structural controls on the new media forms, it will be imperative for the courts to embrace the First Amendment print model standard and apply it to the new forms."[2] In other words, attempts in this country to distinguish rules governing broadcasters—such as the laws prohibiting indecent speech on radio and television—are an unproductive "burden."[3] By supporting such print-broadcast dichotomies, the courts can be seen as failing to preserve "core values of the First Amendment."[4]

Core First Amendment Values

Serious questions must be raised about the denial of free speech when it comes to the category of broadcast indecency. Much like obscenity law, which Professor Thomas Emerson once observed did not follow "most of the rules developed in other areas of the First Amendment," indecency law is an anomaly.[5] In the case of obscenity, "[t]his state of affairs is probably due in large part to the intense and emotional pressures on the courts from the conventional wisdom which views obscenity, at least when available to others, as highly corrupting of the mind and spirit."[6] Emerson argued for drawing a solid "distinction" between expression and action,[7] for providing *different* First Amendment protections for children,[8] and for "sensitive tools" of law "for bringing the basic standards for obscenity regulation more fully into line with the requirements of a system of free expression."[9] For one, questions of prior restraint are at issue under any system that sets out to prevent a category of speech—such as broadcast indecency. For another, one must distinguish between the protection-of-children argument about short-term and long-term harmful effects and the more original concern over impact on moral standards.[10]

Freedom can be seen as a balance between classical liberalism ("doing what one chooses to do in the absence of restraint") and an individual "perfectibility ("the 'ought' and the perfect self").[11] The key, then, is the government's "tolerance" for "surplus repression." In the end, liberty is seen to have a relationship with truth: "[T]here is an objective truth which can be discovered, ascertained only in learning and comprehending that which is and that which can be and ought to be done for the sake of improving the lot of mankind."[12] However, one discovers truth, in this view, "not through intuition and insight, but through a delicate process of interpretation, evaluation, and synthesis."[13] In an ideal world, individual speakers might rise above social limitations, and audience members might be free to sift through and separate right from wrong in a disicriminating process. While this does not seem possible under normal circumstances, most free-expression advocates ultimately come down on the side of tolerance over controls.

At least one author classifies obscenity as "entertainment speech."[14] It is clear that the broadcast indecency case law does share a legal history with obscenity concerns over adult bookstores, movies, and the access by children to them. The distinguishing characteristic of broadcast indecency seems to be that words are at the heart of it. So, comedians such as Lenny Bruce, Mort Sahl, and George Carlin used words in social criticism to break down taboos.[15] Nat Hentoff writes that Bruce's idea was "that if people didn't use language to conceal from themselves what they actually do and want to do, life would be a lot more open and flowing."[16] These ideas led to a 1964 charge in New York for public obscenity for using the words "cocksucker, tits and ass, fuck and other words" in his comedy club act.[17]

What seems absent from our thinking about indecent speech—which in broadcasting seems to include indecency, obscenity, and profanity—is a way to understand the First Amendment as a tool to protect speech. A core question is: Is broadcast indecency an unprotected kind of speech?[18] The answer seems to be that it is "prohibitable" at hours of the day when it might be assumed by regulators to "harm" children. Slightly more protected than obscenity or pornography, time of day indecency restrictions treats it as less valuable than core valued speech: political, social, religious, asthetic, and commercial.[19]

Comparative Issues

Beyond the concern over American law in a global context, comparative issues must be raised. There are no guarantees. The future global standards could be either more or less permissive than existing American law. The "new world" of global communication might not be bound by legal, historical tradition—particularly in a technological age that pays little homage to tradition.

American researchers are still in the early stages of recognizing that failure to observe communication in a global context has blinded our collective interpretations. An international approach "helps us see our communication arrangements in a fresh light, enriches the raw material sources . . . and deepens our appreciation of communication policy issues, learning how they have arisen and been dealt with in other places and periods."[20] Academics who are "cutting edge" enough to explore comparatively the U.S. system with others, must recognize we are a long way from a worldwide communication system: "[E]ven if and when a world system should be common to all of us, we would no doubt keep the regional, national, and local symbol systems of our own for a long time to come."[21]

Media futurists have focused on technology, while entertainment of the masses is likely to remain a key function of mass media: "Entertainment's proven audience appeal acts as bait attracting investment in new technology—what some have called 'digital entertainment.' "[22] The working assumption is that most broadcast systems, previously under the tight control of authoritarian governments, will move toward the American model: "By the turn of the century, most media systems around the world will look more like ours than they do now—entertainment driven and advertiser supported, operating as businesses rather than as arms of government."[23]

This is not to say, however, that the brave new world will be free of governmental intervention and economic/cultural protectionism: "There is far too much at stake—economic power, employment, social and cultural self-image—for most nations to submerge their systems and industries (and, to a large extent, their very identities) in a worldwide media market."[24] The extent to which a Howard Stern, in particular, or broadcast indecency in general could survive abroad will depend upon cultural values about such matters as sexuality, male and female roles, and the interpretation of protection of children and/or innocents.

Rules Governing Broadcast Language in a Global Context

First, it must be understood that the world offers the business of American broadcasting a huge marketplace. "More than 150 countries have radio broadcasting, with close to a billion sets available."[25] A global media environment, however, would open American broadcasting to competition—some of which might also violate the FCCs current indecency standards. Control and ownership patterns in broadcasting vary from country to country. Where capitalism is favored over socialism, technologies such as direct-broadcast satellites threaten to "violate national sovereignty" as "foreign values" might "replace indigenous value systems."[26]

In one sense, these new global issues are similar to what some American communities have faced as they argue national network programming does not meet local community standards of decency. For example, a Utah group failed to ban cable TV nudity when the courts ruled a state obscenity statute could not be applied.[27] Because of the development of nonbroadcast technologies, it is unclear how the continued regulation of indecency can be accomplished. At best, authors such as Orlik (1992) attempted to identify key questions:

- Should basic cable, given its wide access, be subject to the same restrictions as broadcasters while pay services, due to their separate and voluntary purchase, remain exempt?
- What about DBS and MMDS (wireless cable) transmissions?
- More fundamentally, does Congress or the FCC have the constitutional right to regulate the content of nonbroadcast signals?

If not, then why should the federal government continue to restrict content aired on the broadcast stations that must compete with these other services.[28] Orlik quotes J. Brian DeBoice, a communication lawyer, who argues that "broadcasters stand guard at the first breach in the wall of First Amendment freedoms" in the indecency cases.[29]

Realities of Policy Outside the United States

Although we rarely hear about it, broadcast indecency is a worldwide issue. The regulation of broadcasts overseas is not unlike the American situation. Consider this European newswire account from December 1994:

Richard Branson's Virgin Radio has been fined £20,000 by a radio watchdog, for breaching taste and decency rules, the station said tonight. The fine represents the third time the station has fallen foul of the Radio Authority and was imposed after Nick Abbot's late-night phone-in show invited callers to share their sexual fantasies on-air. One man telephoned the station to explain how he enjoyed urinating on his partner during sex. A spokeswoman told PA News that Abbot was fined £1,000 once before, by the station itself, for crudity during a phone-in. On that occasion he was also suspended. Now the troubled DJ, who has worked at Radio Luxembourg, GLR, and at the Virgin Megastore, presents a music-only show. Virgin Radio chief executive David Campbell said: "We fully support the Radio Authority decision. We hope the Authority continues to take an aggressive stance against indecent broadcasts—they have no place on commercial radio."[30]

One element not seen in the American model of regulation is the imposition of a fine on a performer by the station. Where American broadcasters have not stood by their talent, a suspension or firing—not fines—usually follows.

Broadcast Indecency in the Future

Consider the case of Dr. Judy Kuriansky—a Ph.D. whose syndicated show *Love Phones* was filled with frank talk and humor about sex. Robert Santiago of the *Cleveland Plain Dealer* in 1994 argued that "[b]y setting up camp between the prurient and the puritan, *Love Phones* triumphs as redeeming social value and solid entertainment."[31]

If such a program is "sex education with a laugh track," then broadcasters have found a loophole in the current regulatory scheme. Informed talk about sexuality, grounded in the credibility of a knowledgable host, offers the opportunity to appeal to audiences without fear of indecency violations. "The man who stimulates himself with objects, the woman who sleeps with men she barely knows, the couple whose sex life has lost its passion—they are all searching for romantic and nurturing love, Kuriansky says," writes Santiago. Even in this venue, however, the regulatory threats have not vanished: "Producer Sam Milkman screens all callers and makes sure that no objectionable words get aired across the fruited plain. The Federal Communications Commission doesn't mind if you talk about intercourse, as long as you don't use four-letter words," Santiago writes.

Talk radio appears to press the limits of current regulatory tolerance. But Nat Hentoff argues: "The commission might pay heed to D.C. Circuit Judge Patricia Wald, who says of the indecency rule that the FCC should let parents decide what children watch instead of being the nation's nanny. Also, why should adults, by order of the FCC, be limited—until midnight—to what children can hear?"[32] One commentator argued that the language is a sign of something larger:

> "Offensive language is so common among young people that it seems to have lost all of its value as a sign that something genuinely significant was being said. Some comedians use it as punctuation, to squeals of delight from audiences. It is uttered thoughtlessly and with depressing regularity, a great waste of wordage and at the same time a sign of the crack in the framework of public discourse."[33]

Madigan argues: "What everyone seemed to be missing as these broadcasting debates ground on was that the language in question was already embedded and spreading in society, radio or not."[34]

We are left, then, with a regulatory system that sometimes has difficulty with the social context of mass mediated communication. Regulating "decency" on the Internet and other new technologies creates new problems. Speech will continue to evolve in the future, and broadcast speech will do the same. A First Amendment, free speech orientation *must* favor maximum tolerance for broadcast and other forms of indecency—especially where the evidence fails to show negative effects on children. At the same time, an enlightened approach by broadcast managers would be to urge a higher standard for the speech conducted over the public airwaves. Because a major justification for free speech is the search for truth and progress, we need to be proponents for speech that moves us forward.

Notes

1. Jonathon W. Emord, *Freedom, Technology, and the First Amendment.* San Francisco: Pacific Research Institute for Public Policy, 1991, p. 308.
2. Ibid., p. 310.
3. Ibid.
4. Ibid.
5. Thomas I. Emerson, *The System of Freedom of Expression.* New York: Random House, 1970, p. 467.

6. Ibid.
7. Ibid., p. 496.
8. Ibid., p. 497.
9. Ibid., p. 515.
10. Ibid., p. 499.
11. Joyce Flory, "Implications of Herbert Marcuse's Theory for Freedom of Speech," in *Perspectives on Freedom of Speech*, Thomas L. Tedford, John J. Makay, and David L. Jamison, eds. Carbondale: Southern Illinois Press, 1987, pp. 77–89.
12. Ibid., at p. 85, quoting Marcuse.
13. Ibid.
14. Darien A. McWhirter, *Freedom of Speech, Press, and Assembly*. Phoenix: Oryx, 1994, pp. 59–81.
15. Nat Hentoff, *Free Speech for Me—But Not for Thee*. New York: HarperCollins, 1992, p. 322.
16. Ibid., p. 323.
17. Ibid., p. 324.
18. See William W. Van Alstyne, *Interpretations of the First Amendment*. Durham, N.C.: Duke University Press, 1984, p. 38.
19. Ibid., pp. 41–42.
20. Jay G. Blumler, Jack M. McLeod, and Karl Erik Rosengren, "An Introduction to Comparative Communication Research," in *Comparatively Speaking: Communication and Culture across Space and Time*. Newbury Park, Calif.: Sage, 1992, p. 3.
21. Ibid., p. 286.
22. Sydney W. Head, Christopher H. Sterling, and Lemuel B. Schofield, *Broadcasting in America: A Survey of Electronic Media*, 7th ed. Boston: Houghton Mifflin, 1994, p. 572.
23. Ibid., p. 577.
24. Ibid., p. 578.
25. John C. Merrill, John Lee, and Edward Jay Friedlander, *Modern Mass Media*, 2nd ed. New York: HarperCollins College Publishers, 1994, p. 413.
26. Ibid., p. 415.
27. Lynne S. Gross, *Telecommunications: An Introduction to Electronic Media*, 4th ed. Dubuque, Iowa: Wm. C. Brown, 1992, p. 442.
28. Peter B. Orlik, *The Electronic Media*. Boston: Allyn and Bacon, 1992, p. 455.
29. Ibid., citing J. Brian DeBoice, "Monday Memo," *Broadcasting*, 25 August 1989, p. 25.
30. Jonathan Brown, "Virgin Fined after Indecent Sex Phone-in," *Press Association Newsfile*, 16 December 1994.
31. Roberto Santiago, "Sex Show has Cleveland Talking," *Cleveland Plain-Dealer*, 1 December 1994, p. 12E.
32. Nat Hentoff, "The FCC and the First Amendment," *Washington Post*, 26 November 1994, p. A-23.
33. Charles M. Madigan, "America's foul mouth, It's getting worse: Does anybody care?" *Chicago Tribune*, 2 October 1994, P. 12; zone: C.
34. Ibid.

Appendix A

Letters of Complaints

FEDERAL COMMUNICATIONS COMMISSION
WASHINGTON, D.C. 20554

October 26, 1989

IN REPLY REFER TO:

8310-TRW

CERTIFIED MAIL—RETURN RECEIPT REQUESTED

Infinity Broadcasting Corporation, Licensee
Radio Station WXRK-FM
600 Madison Avenue, Fourth Floor
New York, NY 10022

Dear Sir or Madam:

This is in reference to a letter received by the Commission complaining- of the broadcast of allegedly indecent language by station WXRK-FM, New York, NY. More specifically, this complaint alleges that the Howard Stern Show, which is broadcast from 6:00 a.m. to approximately 10:00 a.m. on December 16, 1988, broadcast material which may have violated 18 U.S.C. Section 1464. Enclosed is an example from transcripts submitted by the complainant which the Commission believes may indicate that the subject broadcast contained indecent material.

Section 1464 provides: "Whoever utters any obscene, indecent or profane language by means of radio communication shall be fined not more than $10,000 or imprisoned not more than two years, or both." The Commission has statutory authority to take appropriate regulatory action when licensees broadcast material in violation of Section 1464. 47 U.S.C. Sections 312(a)(6), 503(b)(1)(D).

In three cases concerning broadcast indecency, decided on April 16, 1987, and which we affirmed upon reconsideration on November 24, 1987, the Commission stated that in enforcing the provisions of 18 U.S.C. Section 1464 it would apply the definition of indecency previously affirmed by the Supreme Court in FCC v. Pacifica Foundation, 438 U.S. 726 (1978). See Pacifica Foundation Inc., 2 FCC Rcd 2698 (1987), The Regents of the University of California, 2 FCC Rcd 2703 (1987), and Infinity Broadcasting Corp. of Pa., 2 FCC Rcd 2705 (1987), order on reconsideration, Infinity Broadcasting Corp., 3 FCC Rcd 930 (1987) ("Reconsideration Order"), aff'd in part and remanded in part sub nom. Action for Children's Television v. FCC, 852 F.2d 1332 (D.C. Cir. 1988) ("ACT I"). Under this definition, indecent material is "language or material that, in context, depicts or describes, in terms patently offensive as measured by contemporary community standards for the broadcast medium, sexual or excretory activities or organs." Such material may not be broadcast when there is a reasonable risk that unsupervised children may be in the audience. See generally ACT I. Although a statutory 24-hour ban on the broadcast of indecent and obscene programming was recently enacted, implementation of that statute has been stayed by the D.C. Circuit Court pending its review on the merits of the ban. Action for Children's Television v. FCC ("ACT II") D.C. Cir. Case No. 88-1916. Accordingly, the D.C. Circuit's decision in ACT I currently is the governing law.

FIGURE A.1

Based on the facts furnished by the complainant, it appears that the broadcast of the material in question by station WXRK-FM may have violated Section 1464 by including indecent programming aired during daytime hours. In particular, it appears that the material broadcast is "clear and capable of a specific, sexual meaning" and is patently offensive. <u>Infinity Broadcasting Corp. of Pa.</u>, 3 FCC Rcd at 933. In order to assist the Commission in its evaluation process, we direct you to respond, within 30 calendar days from the date of this letter, to the issues presented herein. In addition, it is our understanding that the referenced material may have been broadcast on stations WYSP-FM, Philadelphia, PA., and WJFK-FM, Manassas, VA. Please advise whether or not this material was broadcast on stations WYSP-FM and WJFK-FM and the times and dates of such broadcasts.

Your response will be used to assist us in assessing whether any violation has occurred. If, on the basis of the information in the record, we find a violation, appropriate action will be forthcoming. Failure to respond to this letter within the 30-day period will result in a violation of Section 73.1015 of the Commission's Rules and a possible sanction.

Sincerely,

Edythe Wise, Chief
Complaints and Investigations Branch
Enforcement Division
Mass Media Bureau

Enclosure

FIGURE A.1 *Continued*

SIGNIFICANT EXCERPTS from "Christmas Party"
broadcast on <u>WXRK</u> <u>New York</u>,
December 16, 1988—from about
8:00 to 10:30 A.M. (Transcribed
from magnetic tape recording
made on personal Panasonic by
Anne Nelson Stommel.)

<u>Side 1</u>
inches (approx)

017—   To tune of "Jingle Bells." Howard Stern
        says, "What, did he sample one of his farts
        into a Casio?"

060—   Now Bo (the Lesbian of course) is going to
        be serving drinks, <u>naked</u>. Don't give me a
        no. I don't want to hear a "no". You
        decided you were going be nude and you've
        got to stick to it.

065—   (To serve drinks) Bo you're going to need a
        tray. Or, you can use your breasts . . . And
        she does sleep around a lot.

160—   'Gina Girl, what are your going to be doing
        for us today? ('Gina Girl is short name for
        Vagina Girl.)

167—   You know, Christmas music is really
        maudlin. You wouldn't think the baby Jesus
        was born. You'd think somebody was abusing
        him.

245—   Hey Doc (hypnotist). Can you do it so that
        when I rub her ear Robin has an orgasm? You
        can do that? I rub <u>my</u> ear?

266-287 And, 'Gina Girl, you work at a Retarded
        Place? . . . She's taking her clothes off.
        Wait 'til the people in back get a whiff! I
        don't know. Suddenly I'm nauseous and never
        want to see another naked woman.

287-311 Get her clothes back on . . . Oh, my God
        . . . I don't know. This show's going all to
        Hell . . . Now I think I've seen everything
        . . . Wow . . . 'Gina Girl is wild . . . The

FIGURE A.2

lesbian wants a guy now . . . The rest of the normal girls will take their clothes off . . . Some guys are throwing up . . . I don't know how to describe this . . . I've never seen men more turned off. Vinnie Mazzie who lit his penis on fire and eats crickets is throwing up.

317-   I told you we should have filmed this for a TV Special. 'Gina Girl, do you realize everybody's repulsed? I have never averted my eyes from a naked woman before in my life.

328-331 You couldn't even tell us what kind of breasts she had. I don't know. I didn't look. I saw her belly hanging over.

Side 2
inches (approx)

009-   We're back at the Christmas Party . . . and I gotta tell you it's wild in here, Robin . . . The guy who plays the piano with his wiener is here now.

017-041 (Gay choir) We got two gay guys and a heavy-set woman lesbian (Negro). Remember you're on the radio, will you honey? I get called a fag hag. Have you ever had a man? Have you been with a man? Disappointment, hell. Oh, you like it? Well, it's just not a preference. It just doesn't turn me on as much . . . You gotta be glad about the 5-minute AIDS test. Now you guys can test each other and then hop into the sack.

066-   What is it that men don't find me attractive?... men who find other men attractive... my uh? ... Your small penis probably ...

079-   How about this? "A Tuckis so Bright"?

090-   (Gay choir) - "I'm dreaming of some light torture, some bruises just to make me moan ... Masturbate. Humiliate. Gay sex is fun

FIGURE A.2   *Continued*

in the city. Howard Stern is going to learn how great a tuckis can be ...

196-198   (To Robin, about to be hypnotized - by Dr. Marshall King) Just think about this. Every time I rub my ear, you'll be orgasming.

231-   Vinnie Mazzie, ladies and gentlemen. The man who lit his penis on fire ... and now feels that he can eat six live crickets ...

Side 3
inches (approx)

058-   Here's a guy who plays the piano with his penis. I'll tell you. What a weird crew!

077-   She doesn't give me anything as far as jewelry is concerned. That's because you don't have a penis.

114-   I think it will be worth the whole show just to hear her have an orgasm when I rub my ear.

179-186   I understand the doctor (hypnotist) explained that you would not go for that orgasm thing. Like, in other words, that you would not accept it... Boy you must be difficult in bed. He says he'll hypnotize Bo the Lesbian ... and she's a good subject. She's an empty slate.

221-   Bo, you look great. Yes ... Bo ... getting very aggressive. Bo just rubbed herself in my face. Juliet (one of naked girls) getting wild. Oh, my God. Diane (another naked girl) is whipping Bo. Good, girls, excellent. The big black lesbian is out of her mind with lust. Look at her. You can't say it on the air? Were you getting excited? Fabulous. All right. That was really good. Best part of the whole Christmas Party.

253-269   I want to rub my ear and have this girl go wild for me ... When we came back from commercial, we have a young man who wants

FIGURE A.2   *Continued*

to play the piano with his, uh, wiener ...
Howard, I'd better go into the other room
and, uh, get it ready. I'll come strolling
in swinging it. It's bigger than yours.
I've got a rubber. Don't worry about it.
And I've got a second rubber for an encore.
He's going to wear a contraceptive. I do
safe organisms ... orgasms. I'm going to
play the Casio ... I believe we hit two
keys at the same time. You'd better give me
the next segment, though. I'm going to get
it going. O.K? Go in the other room and do
whatever you have to do to play it.

270-283 The doctor is now hypnotizing Bo and he
wants to know exactly what it is you want
her to do. The orgasm thing? I want more. I
was thinking of something a little heavier
like beg me to make love to her. Oh,
there's an idea. You know what I mean? That
would be a change for you. I want to rub my
ear and have her beg me ... go wild for me
... get off her chair ... she can't resist
me. She'll come and attack you. Right.
That's what I want ... I've never had that.
Let's see the Doc do that.

331-341 We'll be back right after this with a
hypnotized Bo and the guy who plays the
piano with his penis . . . It's a <u>Christmas</u>
Party! . . . More gay Christmas Songs . . .
and the burper . . . he's going to be
belching for us . . . and, um . . . 'Gina
Girl might even be persuaded to do the
'Gina Dance again. Is there a chance you
could do the 'Gina Dance? . . . 'Gina Girl
did you vote in the last election? Yes. Oh,
you did? Who did you vote for? . . . Bush.

<u>Side 4</u>
inches (approx)

003- All right, bring her in. You're saying that
she'd actually crawl over me? I just think
that she'd think she was in bed. Leap over.

FIGURE A.2   *Continued*

How're you doing Bo? So, the doctor (hypnotist) and Bo are back. This is great. You know I've got the itchiest ear. Look at this. Why are you doing that? Doing what? Stop, please! Oh, my God! This doctor's amazing. I'm telling you. Oh, my God! I love having this power over women. It's so fabulous . . . Bo, do you find me attractive?

029- She's really diggin . . .

034- Doc, you're going to do something else? Bo, when you wake up, you see the counter in front of you? and Howard . . . he's got his thing right up on the counter . . . and still when he scratches his ear it still affects you . . . but it's right up here on the counter . . . Do you want to touch it? Go ahead. You can touch it. Who is the biggest man in the whole world? Howard. Am I your hero? Yes. Now you'll worship me? Oh man! My dream come true! <u>What</u> <u>a</u> <u>Christmas</u>! This is good. Tony, I really like this.

054- I like the sexual stuff better myself.

073- She just sat there and got quivery. . . . I want her to be <u>totally</u> crazed for me. The doc's going to do it. It will be very difficult for you to keep yourself away from Howard. You want to be near him. You want to touch him. Nobody's in the room. You don't give a damn who's in the room. You want to be next to Howard. You want to hold him, touch him. No one can take him away. And if another girl even tries to get near, you can get very nasty.

(NOTE: This goes on for 40 more inches. Bo wants to tear off Howard's shirt. Some of the words: Go lower. What is it you see when my shirt is off like this? Potential. I just want you . . . They're fighting over me. This is heaven. Why can't I get my wife to act like this?)

FIGURE A.2 *Continued*

140–   I just realized. They should have had me and Juliet in that hypnotism thing. Then I would really have been out of my mind.

171–   What is next, Gary? Well, um, we have some trouble. Yes. Our General Manager's threatening to put an end to this party. Why? Because, um, some of the participants have been sneaking out and doing things they shouldn't be doing. And I can't get into it, but it's really gross. Well, tell that General Manager to shut up. No, no, no, he's right on this one. Its not going out over the air. I know, but I guess he doesn't like it. He doesn't want it in the office. He doesn't want it in the office? Right! Well, that's <u>his</u> problem. Tell him to deal with it. What are they doing out there? You can't say it on the air? It's like the third floor of Bloomingdale's Men's Room. O.K? Oh, really! Wow! Oh, man. Who's not in the room?

181–   We haven't got to Robin's stripper yet. We haven't got to the piano guy. Right. Well, I don't think we're <u>going</u> to be getting to him. And then we want to see the girls nude again.

164–   The piano guy's ready now . . . like at this second. Oh, he <u>is</u> ready. O.K. bring him in. Robin, I think you're going to be amazed by this . . . our <u>Christmas</u> Party . . . Bo, you were really diggin me . . . All right, here he is. Here he goes. (Plays some notes of Jingle Bells, ending with a flourish.)

192–   I don't know how to play this damn thing . . . I'm tired anyway . . . Look at the gay guy. He's going wild . . . All right, <u>very good</u> . . . Let me wipe that thing off. Gee, I'm not playing <u>that</u>! Yeah, take it home with you. I could have worn my rubber. But, you know, I could have never done it. You know?

FIGURE A.2   *Continued*

203– We'll be back with more of the <u>Christmas Party</u> right after these words.

215– (Organ music—"I'm Dreaming of a White Christmas") All right. We're back. We're back at the <u>Christmas Party</u>. I'll tell you. It's a lot of fun. It really is . . . I think my favorite part so far, though, was when Bo was hypnotized. You have no idea what happened to you? . . .

219– I loved it. When you touched your ear? Yes. It was the most incredible feeling I've ever felt. It was just <u>so great</u>! Yeah, really? I mean I couldn't help it. Hey, it was so erotic. Wow! I mean I wished you wouldn't stop. My God. It was so good. Oh, my God, I wanted it so bad. Really? Remember sitting on my lap and attacking me? I just wanted to get at you right <u>here</u>. I was so angry when the girls came over. Angry, I don't know why . . .

229– God, our General Manager got upset because two of our participants in the Howard Stern Birthday (sic) Show were participating in homosexuality in the bathroom. But I gotta tell you something . . . it's not <u>my</u> fault. I mean, I hope he's not mad at <u>me</u>! I can't control that . . .

230– Now, Bo, does any of that residual stay with you? . . . No, it's still lingering. Really? You mean if I rub my ear you could still do it? Yeah . . . I mean not as much . . . but I feel a little faint memory. You do! Yeah. . Wow! That is <u>awesome</u>! That doctor. We should have had him hypnotize all the girls. . . . It was really good. It's just a shame we couldn't finish it . . . Well, I don't know. You stopped. It was like ripping my shirt off made you stop. I didn't know if the bottom was taboo for some reason. It was, eh? Because he didn't tell you. Yeah. I couldn't go beyond.

FIGURE A.2 *Continued*

250–   What happened when you wanted me so bad?
       Was I like the best looking guy you ever
       saw? Ye-ess. Incredibly sexy. It was better
       than . . . You <u>saw</u> it? Right. Much better
       than you saw it. You tell her she was
       hypnotized. All right? It was pretty good.
       I don't know why he put it on the table,
       but . . . I just wanted you to see it for
       once . . . Oh man, that was wild. It was
       great. You're still looking at me like you
       love me. I know, déjà vu. Really! You can
       still look at me and get excited . . . Yeah.
       Oh, that's great.

267–   Well, the party's really getting wild now.
       The condoms have been blown up and people
       are passing them around like balloons.
       They've only been used twice, Howard. Thank
       you . . . What was in there? I just opened
       it. Here it is. It's probably the stuff
       that comes inside. Oh, no. It's kinda wet.
       Oh, no. I wouldn't want <u>you</u> in there. . . . I
       feel like I should spend the rest of the
       holiday in church . . . On your knees and
       someone's passing condoms around . . . This
       is, um, the <u>worst</u> Christmas Party we've
       ever done. I mean, is this like the most
       <u>sinful</u> Christmas Party? Debauchery. This is
       the end of the Roman Empire. It really is
       . . . Hey, I think it's time for another gay
       Christmas Carol . . . What do you think of
       this Christmas Party? . . . It will be hard
       to top <u>this</u> one. . .

315–   How about the gay Christmas Choir, now?
       What are you guys going to sing? Its a song
       . . . it's a version of "Winter Wonderland."
       (Many words not clear): In the dark by the
       _____ a beautiful night with tuckis so
       tight, Christmas at the _____ as I watch a
       tuckis gyrate, take my manhood to make
       _____ good, Christmas at the _____. We can
       dress you like a gladiator, make believe
       that you're a sweaty jockey or pretend that

FIGURE A.2   *Continued*

you're a _____ Mary and fantasize that you
are Liberace. With a vice and some winches
we'll stretch you to fourteen inches. We've
_____ you so strong and _____ go wrong,
Christmas at the _____. All right! Very
beautiful. Lovely . . . Are we missing
anyone? The dancers. The male dancers.

381-   I know I'm going to Heaven. Because God
forgives me because I _____ Video Smarts.
This is for children 3 to 6 years old. Good
children. Good, solid children. Children
like _____. Yeah, I mean how would you like
your children to grow up taking his penis
and playing the piano with it? Or, how
would you like her doing the 'Gina Dance?
That's because these children didn't have
Video Smarts . . . How would you like her to
grow up to be the Gay Choir? . . . This is
what I'm talking about.

390-   Or the burper. I don't think he was that
good. I don't think we <u>let</u> him. Well, let's
hear it now. Come up to the microphone if
it's so damn good. (Belch, belch, belch)
You want "Jingle Bells"? (Belches out the
beginning notes of Jingle Bells.)

NOTE: During a mid-portion of the program, Robin had
said: You know what I was just thinking? I really
feel bad that 65-year-old woman isn't here today.
Maybe we should call her <u>while</u> they're taking their
clothes off. At the end of the program, Stern phones
Philadelphia and Washington to ask how they liked
the Christmas Party. Then Stern says: Maybe we
should call that old woman who hates us. Stommel had
telephone tape answering machine on throughout radio
program but did not intend to answer phone if WXRK
should call. Following is recording of end of
"Christmas Party" tape: Click . . . Buzz . . .
Recorded Voice: "Your call cannot be completed."
Robin: Don't tell me she changed her number already.
Buzz, buzz, buzz. Stern: She might have. It's
possible, Robin. I don't know. I don't think she

FIGURE A.2   *Continued*

could do it that quick. Because I've changed my
number many times. Robin: Yes, but in the holiday
rush season (Ring, ring) I don't think the telephone
company is going to be that quickly. (Ring, ring)
Background Voice: "Hello, This is Anne Stommel.
Please don't hang up just because this is my tape
answering maching. . . ." Stern: Maybe she passed out
from the . . . Robin: Yeah, I was going to say Anne
must have passed out. Background Voice: "If you'll
leave your name and phone number . . ." Oh, that was
a tape. Stern: We'll have to try again . . . maybe
next Christmas. . .

FIGURE A.2   *Continued*

Infinity Broadcast Corp.
Radio Station WXRK-Fm, New York, NY
The Howard Stern Show
6:00-10:00 a.m.

December 16, 1988

We're back at the Christmas Party...and I gotta tell
you it's wild in here, Robin...The guy who plays the
piano with his wiener is here now.

(Gay choir) We got two gay guys and a heavy-set
woman lesbian (Negro). Remember you're on the radio,
will you honey? I get called a fag hag. Have you
ever had a man? Have you been with a man?
Disappointed, Hell. Oh, you like it? Well, it's not
just a preference. It just doesn't turn me on as
much...You gotta be glad about the 5-minute AIDS
test. Now you guys can test each other and then hop
in the sack.

What is it that men don't find me attractive?...men
find other men attractive...my uh?...Your small
penis probably...

How about this? "A Tuckis so Bright"?

(Gay choir) - "I'm dreaming of some light torture,
some bruises just to make me moan...Masturbate.
Humiliate. Gay sex is fun in the city. Howard Stern
is going to learn how great a tuckis can be...

(To Robin, about to be hypnotized by Dr. Marshall
King) Just think about this. Every time I rub my
ear, you'll be orgasming.

Vinnie Mazzie, ladies and gentlemen. The man who lit
his penis on fire...and now feels that he can eat
six live crickets...

Here's a guy who plays the piano with his penis.
I'll tell you. What a wierd crew!

She doesn't give me anything as far as jewelry is
concerned. That's because you don't have a penis.

I think it will be worth the whole show just to hear
her have an orgasm when I rub my ear.

FIGURE A.3

I understand the doctor (hypnotist) explained that
you would not go far that orgasm thing. Like, in
other words, that you would not accept it...Boy, you
must be difficult in bed. He says he'll hypnotize Bo
the Lesbian...and she's a good subject. She's an
empty slate.

Bo, you look great. Yes...Bo...getting very
aggressive. Bo just rubbed herself in my face.
Juliet (one of the naked girls) getting wild. Oh, my
god. Diane (another naked girl) is whipping Bo.
Good, girls, excellent. The big black lesbian is out
of her mind with lust. Look at her. You can't say it
on the air? Were you getting excited? Fabulous. All
right. That was really good. Best part of the whole
Christmas Party.

I want to rub my ear and have this girl go wild for
me...When we come back from commercial, we have a
young man who wants to play the piano with his, uh,
wiener...Howard, I'd better go into the other room
and, uh, get it ready. I'll come strolling in
swinging it. It's bigger than yours. I've got a
rubber. Don't worry about it. And I've got a second
rubber for encore. He's going to wear a
contraceptive. I do safe organisms...orgasms. I'm
going to play the Casio...I believe we hit two keys
at the same time. You'd better give me the next
segment, though. I'm going to get it going. O.K? Go
in the other room and do whatever you have to do to
play it.

The doctor is now hypnotizing Bo and he wants to
know exactly what it is you want her to do. The
orgasm thing? I want more. I was thinking of
something a little heavier like beg me to make love
to her. Oh, there's an idea. You know what I mean?
That would be a change for you. I want to rub my ear
and have her beg me...go wild for me...get off her
chair...she can't resist me. She'll come and attack
you. Right. That's what I want...I've never had
that. Let's see the Doc do that.

We'll be back right after this with a hypnotized Bo
and the guy who plays the piano with his

FIGURE A.3   *Continued*

penis...It's a <u>Christmas</u> Party!...More gay Christmas
Songs...and the burper...he's going to be belching
for us...and, um...'Gina Girl might even be
persuaded to do the 'Gina Dance again. Is there a
chance you could do the 'Gina Dance?...

FIGURE A.3   *Continued*

Appendix B

Station Response Letters

REED SMITH SHAW & McCLAY
INCLUDING

PIERSON, BALL & DOWD

1200 18TH STREET, N.W.
WASHINGTON, D.C. 20036

202-457-6100

FAX 202-457-6113
TELEX NO. 64711

WRITER'S DIRECT DIAL NUMBER

(202) 457-8647

PITTSBURGH, PA
PHILADELPHIA, PA
HARRISBURG, PA
McLEAN, VA

September 25, 1989

Ms. Edythe Wise, Chief
Complaints and Investigations Branch
Enforcement Division
Mass Media Bureau
Federal Communications Commission
2025 M Street, N.W., Room 8210
Washington, DC 20554

> Re: 8310-TRW
>   Radio Station KSJO-FM
>   San Jose, CA

Dear Ms. Wise:

On behalf of Narragansett Broadcasting Company of California, Inc., the licensee of radio station KSJO-FM, San Jose, CA, we are submitting herewith the licensee's response to your letter dated August 24, 1989 in the above-referenced matter.

Please contact this office if further information is needed.

Very truly yours,

REED SMITH SHAW & McCLAY

Peter D. O'Connell

PDO/jk
Attachment

FIGURE B.1

Ms. Edythe Wise, Chief
September 21, 1989
Page 2

Letter eight examples of material from the Perry Stone Show
submitted by the complainants.[2]

KSJO did not retain recordings of the materials cited by
complainants, and therefore we are unable to comment on the
accuracy of the transcripts cited. We assume that tape
recordings have been submitted to the FCC. Accordingly, for
purposes of this response we presume that the transcripts are
correct.

Before addressing your specific question, we wish to note
that Perry Stone was suspended by KSJO on March 14, 1989 and then
fired on March 21, 1989. He had violated station operating
policies and guidelines, and generated numerous audience
complaints.

KSJO's failure to take this action earlier was the product
of (1) Stone's repeated promises to conform to our guidelines and
"clean up his act"; (2) his program's strong and extremely loyal
following and support among KSJO's adult listeners; and (3) as
will be discussed further below, genuine uncertainty on the part
of the station management concerning the "contemporary community
standards" that were applicable to Stone's program in regard to
KSJO's radio audience.

The precipitating reason for Stone's discharge was his
unwillingness to accept direction in matters of program taste and
judgment.[3] The station was aware of FCC guidelines on indecent
programming, and had admonished Stone concerning them.

---

[2] The excerpts cited are alleged to be from 1988 broadcasts
carried 7:00-8:00 a.m. on October 20, 25, 27, 28 and
November 4 and 10, and 9:00-10:00 a.m. on November 10 and
14. The Perry Stone Show was carried on KSJO from
December 14, 1987 until March 14, 1989. It was broadcast
Monday through Friday from 6:00 a.m. to 10:00 a.m.

[3] The specific incident that triggered Stone's suspension
involved his improper on-air comments (not sexual in nature)
to two children who were guests on the program, but by this
time KSJO's manager had already determined that Stone was
insufficiently responsive to program direction and could not
be persuaded to satisfactorily conform to KSJO guidelines.
His discharge took into account concern about compliance
with FCC rules and policies (see KSJO's March, 1989 press
release (Exhibit No. 1)).

FIGURE B.1   *Continued*

Ms. Edythe Wise, Chief
September 21, 1989
Page 3

We cite the above facts to make it clear that NBCCI has
taken steps to assure that KSJO will no longer carry the type of
language used by Stone in the excerpts attached to the Staff
Letter. We also regret the occurrence of these broadcasts in
1988.

KSJO's manager during the period in issue was Mr. David
Baronfeld. He reports directly to Mr. John Peroyea of NBCCI.
Mr. Baronfeld assumed his position on August 12, 1988. Since
Perry Stone was already on the air at this time, and was well
known locally as a so-called "Shock Jock," one of Baronfeld's
first tasks was to issue guidelines to Stone on acceptable
programming and language in regard to sexually-oriented humor.

Baronfeld did not monitor every Stone program in its
entirety, but he was completely aware of Stone's approach to
humor. Stated succinctly, Baronfeld understood that his duties,
pursuant to specific instructions from Peroyea, were to keep
Stone's material within acceptable station and FCC bounds of
taste and decency as to sexual matters, without unduly
constricting ribald forms of humor which plainly appealed greatly
to a large and enthusiastic listening audience.[4]

Baronfeld did not hear the specific KSJO broadcasts of the
material cited in the Staff Letter. They were first brought to
his attention by local complainants in November, 1988, and were
reviewed with NBCCI's Peroyea in early December, using tape
transcripts submitted to KSJO by those complainants. NBCCI
directed Baronfeld to take immediate steps to assure that in the
future Stone would not make any sexual references on the air to
bodily parts or functions, and Peroyea reviewed with him in
detail past FCC rulings and notices on indecency. Baronfeld then
met with Stone to reemphasize (orally and in writing) our
policies on these matters. Stone agreed to comply with these
instructions. At the time NBCCI expected these measures to be
sufficient in terms of dealing with the transcript materials
forwarded by complainants.

This background is intended to acquaint the Commission with
some of KSJO's efforts to address the enforcement of station
KSJO's own program guidelines and to apply FCC indecency rulings
in regard to Stone. Your letter presumably seeks our further
response as to whether in NBCCI's view Stone's language, as

---

[4] NBCCI obviously takes full responsibility for Stone's
program and for its employees' supervision thereof.

FIGURE B.1 *Continued*

Ms. Edythe Wise, Chief
September 21, 1989
Page 4

reflected in the October/November excerpts, was "in context"
indecent as a matter of law because it was "patently offensive"
as measured by "contemporary community standards" for the
"broadcast medium."

We do not dispute that the segments cited are tasteless and
susceptible to the judgment that they are offensive to many
listeners. Because NBCCI and a number of persons in our audience
regard them to be so, we sincerely regret that these materials
were broadcast. However, KSJO encountered great difficulty in
judging Stone's program overall when employing the FCC indecency
formulation, because the station was receiving strong local
listener support for Stone (reflected in mail, telephone calls
and audience ratings),[5] and the so-called "standards" for the
broadcast medium are obviously highly subjective in nature and
thus extremely hard to pin down in concrete circumstances.
(Stone often maintained, for example, that he modeled his humor
after a well-known and popular New York City disc jockey who
daily presented very similar kinds of material and has been on
the air for several years.)

The candid answer is that we do not really know if Perry
Stone's humor violated the legal standards to which the Staff
Letter refers, because the "patently offensive" and "community
standards" tests are so vague and essentially subjective that we
do not completely understand how to apply them in specific
instances, especially as to double entendre. The standards,
which are described by the Commission as having industry-wide
application, impose on station operators and their often
irrepressible on-air talent the need to make critical predictive
judgments about how the FCC will characterize national program
values at some indeterminate future date. (In this case, for
example, KSJO was obliged to anticipate the possible judgments of
Commission personnel acting almost a full year after the subject
programs were aired, making it uncertain as to exactly what time
period is encompassed by the reference to "contemporary"
standards.)

Further, it is not at all clear what type of national
information base is used by the FCC in establishing appropriate

---

5   In the last half of 1988 Stone's program was ranked number
     one for its time period for men aged 18-49. His language
     and "shock" humor also had many local critics, of course,
     including the complainants.

FIGURE B.1   *Continued*

Ms. Edythe Wise, Chief
September 21, 1989
Page 5

boundaries for "community" standards in regard to sexually-
oriented material, which by its nature evokes widely varying
personal judgments.

Our admission that Stone had to be repeatedly admonished not
to make graphic sexual references, that we find the excerpts in
question tasteless, and that we had to discharge Stone for
failure to comply with our broadcast policies, do not reflect an
awareness by KSJO that he had in fact "crossed the line" into
violation of the FCC indecency standard. They do confirm that
KSJO attempted to create and enforce compliance with what it
regarded to be contemporary moral and legal standards. Having
failed to gain Stone's compliance, KSJO dismissed him more than
five months before the Staff Letter was received. The program
subsequently carried on KSJO in Stone's time period has elicited
no listener complaints as to offensive sexual language and does
not employ Stone's style of "shock" humor.

KSJO's experience with Stone led it to conclude several
months in advance of the FCC inquiry that his type of humor is
much too difficult to police and too offensive to certain members
of the listening audience to warrant carriage on KSJO. The
station therefore elected to abandon this type of programming
last March, and has no intention of returning to it. In this
regard, KSJO held meetings with certain complainants in May and
June of 1989, publicly apologized to the station listeners who
were offended by Stone's program, and made clear the change in
direction of its morning programming. A joint press release on
this subject was issued on June 30, 1989 (see Exhibit No. 2).

We urge that the Commission also take into consideration the
fact that the state of the law on actionable indecency is still
largely unsettled. Thus, while the Court of Appeals affirmed the
Commission's articulation of an indecency standard, it clearly
showed its recognition of the complexity of applying the
Commission's standard, notably during those times of the day when
significant numbers of children could be expected to be under
parental supervision.[6]

---

[6] Action for Children's Television v. FCC, 852 F.2d 1332 (D.C.
Cir. 1988). The court stated that it was "impelled" by the
Supreme Court's decision in FCC v. Pacifica Foundation, 438
U.S. 726 (1978), to affirm the FCC ruling, but added that
the FCC itself would be acting with "utmost fidelity" to the
first amendment were it to "reexamine, and invite comment
Continued on following page

FIGURE B.1 *Continued*

Ms. Edythe Wise, Chief
September 21, 1989
Page 6

   In view of the factual circumstances we have described, the
ambiguous and unsettled state of the law, and the widely
acknowledged difficulties encountered by many broadcast stations
in interpreting and applying this FCC policy, NBCCI respectfully
submits that no penalties should be levied against it, even if
the FCC concludes that Stone's material was indecent when
evaluated by contemporary standards.

   We are hopeful that this response will assist the Commission
in evaluating the licensee's position and in discharging the
agency's responsibilities in regard to indecent programming.
NBCCI will continue to cooperate with your inquiry in any way
possible, and is prepared to promptly supply such further
information as you may deem necessary.

                              Very truly yours,

                              NARRAGANSETT BROADCASTING
                              COMPANY OF CALIFORNIA, INC.

                              By _____
                                   Gregory P. Barber
                                   President

Continued from previous page
     on, its daytime, as well as evening, channeling
     prescriptions." Id. at 1341. This is a reference to the
     governmental interest in promoting parental supervision of
     children's listening. Id. at 1344. We note in this regard
     that six of the nine cited KSJO broadcasts were aired before
     9:00 a.m., i.e., at times when parents would normally be in
     a position to supervise their children's radio listening.

                FIGURE B.1   *Continued*

Appendix C

Letters of Community Opposition

3320 66th Ave.
Apt. # 3
Oakland, CA 94605
January 27, 1988

Phil Norton
1420 Koll Circle
San Jose, CA 95112

Mr. Norton:

On my way to work the morning of January 4th, I happened to tune into KSJO. It was about 9:30 AM and my first time tuning into the station. I was shocked to hear a song that was to say the least disrespectful, indignant, and insulting. The words of the song included this quote, "My wife ran off with a Nigger". For some reason the D.J. deleted the word Nigger as if by this ommission the song would be more acceptable. This mentality all but amazed me. Its place and its meaning were clearly understood even in its absence.

Any radio station given the privilege of using the public air-waves has a public responsibility to excersise that privilege in the public interest. I feel this principle was clearly violated by the irresponsible act of airing such an offensive song. It was an explicitly racist recording that should never have been aired.

I demand a public apology and a clear statement of your policy regarding this matter. In the absence of these two corrective measures I will have no recourse but to file official complaints with the F.C.C., Urban League, N.A.A.C.P. and the San Jose Chamber of Commerce. You must realize

FIGURE C.1

the racist posture your station assumes by allowing such
unacceptable material to go out over your broadcast and
I feel that it is this type of behavior that could very
well prompt the Federal licensing agency to deny renewal
in the least and revoke at best your broadcast license.

Again, this incident was in poor taste and unacceptably
insulting to anyone concerned with human rights and dignity.

Sincerely

Richard Turner, M.D.

FIGURE C.1   *Continued*

July 19, 1988

Ms. Kay Hickey
KSJO FM
1420 Koll Circle
San Jose, CA 95112

Dear Ms. Hickey,

Your morning disc jockey, Perry Stone, has a hard time distinguishing humor from hatred.

I couldn't believe the "jokes" or "bits", whatever you can call them, that I heard this morning on KSJO. The story about the "Oriental" with the dog, for example. The "Oriental" what? A human? A woman? An animal? It's difficult to tell from the undisguised disgust with which this man attacks the subhuman creature that walks its dog in his neighborhood. Before this story was a truly unfunny, completely racist "comedy routine" about a Vietnamese take-out restaurant in which the punchline is the delivery man is killed by the customer. The Vietnamese people's language is a ridiculous jibberish and of course the foods offered are an assortment of household pets. the story ends with something like "Honey, there's a dead Vietnamese in the doorway." This is some great material!

I take in Alex Bennet's show almost every morning so I'm used to biting and sarcastic humor. But I never felt like reaching for the phone and calling the FCC in Washington like I did this morning. I think I can tell the difference between comedy and racism and your boy's got a big problem with Vietnamese people. By accident my radio was set on your station this morning and I heard 5 minutes of hate. The message your station is giving out to the South Bay is pretty obvious. Vietnamese and Oriental immigrants don't belong in your neighborhood and it's cool to hate them and laugh at them. If you had any skill and a slightly more sophisticated view of the world you'd be figuring out how to laugh with them.

Sincerely,

Wendy Lieber Poinsot
505 Hamilton Avenue, Suite 201
Palo Alto, Californa 94301

FIGURE C.2

Appendix D

Letters of Audience Support

LARRY WERT,

I AM A 22 YEAR OLD STUDENT OF
ARCHITECTURE AT THE UNIVERSITY
OF ILLINOIS, CHICAGO. I THINK STEVE
+ GARY, AND ARCHITECTURE SHARE AT
LEAST ONE THING IN COMMON; THEY ARE
BOTH REFLECTIONS OF SOCIETIES VALUES
AND INTERESTS. AT THE PRESENT TIME
I MAKE LITTLE TO NO DIFFERENCE IN
SOCIETY, BUT WHEN I DO, I HOPE IT
IS WITH THE SENSE OF QUESTIONING
+ REALITY THAT STEVE + GARY POSSESS
+ CONVEY TO THEIR LISTENERS. IN
THEIR OWN LITTLE WAY THEY HAVE
TAUGHT ME TO LOOK, EVALUATE, AND OFTEN
LAUGH AT LIFE AND MORE IMPORTANTLY
MYSELF. BEING A STUDENT IS VERY
HECTIC AT TIMES, AND STEVE + GARY
ARE MY MEDICINE.

M. UNTIEDT
5017 N. HAROLD AVE.
SCHILLER PK, IL, 60176

ALWAYS WILLING TO HELP,

Mark Untiedt

FIGURE D.1

Tony Foley
474 S. York Rd.
Elmhurst, IL 60126

Mr. Larry Wert
WLUP AM-1000
875 N. Michigan Ave.
Chicago, IL 60611

Dear Mr. Wert:

    As an avid listener of Steve Dahl and Garry Meier, I feel compelled to
respond to the flurry of charges which have been leveled against the pair
recently. I am speaking specifically about the alleged "obscenity" of certain
broadcasts. As a lawyer, I fail to see any validity in these charges, but more
importantly, as a member of the general public, I am somewhat amused that some
people aren't smart enough to turn their radio off and that others have enough
time on their hands to monitor every broadcast in a futile attempt to stir up
some sort of trouble. Having grown up in a conservative environment in South
Dakota, one would think that I would be just the sort of person who might find
them to be offensive, but nothing could be further from the truth. While I
certainly would acknowledge that Steve and Garry aren't for everyone (as they
often do themselves during the broadcast), their listeners find the show to be
a refreshing forum where the concerns of an entire generation are openly,
frankly, and more often than not, hilariously discussed. Those who don't care
for the approach of the program are free to search the vast wasteland that is
Chicago radio for an alternative, and I wish them the best of luck, but the
handful of people who may have a complaint about the show should not be
allowed to dictate what the rest of us are able to listen to. To their
thousands of listeners - the only people whose opinions should really count
here - Steve and Garry represent a much-needed oasis during the drive home or
the afternoon with the kids. I commend WLUP for standing firm behind them, and
urge Steve and Garry to hang tough and keep up the good work.

Very truly yours,

Tony Foley
Law Editor, Commerce Clearing House
Hm. Ph. 833-3734
Wk. Ph. 940-4600   ext. 2463

FIGURE D.2

Selected References

Academic Journals and Law Reviews

Atwood, Rita A., Susan Brown Zahn, and Gail Webber, "Perceptions of the Traits of Women on Television," *Journal of Broadcasting & Electronic Media* 30(1):95–101 (Winter 1986).

Barbatis, Gretchen S., Martin R. Wong, and Gregory M. Herek, "A Struggle for Dominance: Relational Communication Patterns in Television Drama," *Communication Quarterly* 31(2):148, 155 (Spring 1983).

Blanchard, Margaret A., "The American Urge to Censor: Freedom of Expression Versus the Desire to Sanitize Society—from Anthony Comstock to 2 Live Crew," Volume 33 *William & Mary Law Review* 741 (Spring, 1992).

Block, Peter Alan, "Modern-day Sirens: Rock Lyrics and the First Amendment," *Southern California Law Review* 63:777 (1990).

Booth-Butterfield, Steven and Melanie Booth-Butterfield, "Individual Differences in the Communication of Humorous Messages," *The Southern Communication Journal* 56(3):205–218 (Spring 1991).

Cantor, Joanne R., "What Is Funny to Whom? The Role of Gender," *Journal of Communication* 26(3):164–172 (Summer 1976).

Carlin, John C., "The Rise and Fall of Topless Radio," *Journal of Communication* 26(1):31–37 (Winter 1976).

Chapman, Anthony J., and Nicholas J. Gadfield, "Is Sexual Humor Sexist?" *Journal of Communication* 26(3):141–153 (Summer 1976).

Donnerstein, Edward, Barbara Wilson, and Daniel Linz, "Standpoint: On the Regulation of Broadcast Indecency to Protect Children," *Journal of Broadcasting & Electronic Media* 36(1):111–117 (Winter 1992).

Dyk, Timothy B., Book Review, 40(1) *Federal Communications Law Journal* 131–141 (1988).

Gey, Steven G., "The Apologetics of Suppression: The Regulation of Pornography as Act and Idea," *Michigan Law Review* 86:1564 (1988).

Goldstein, Jeffrey H., "Theoretical Notes on Humor," *Journal of Communication* 26(3):104–112 (Summer 1976).

Groce, Stephen B., and Margaret Cooper, "Just Me and the B? Women in Local-Level Rock and Roll," *Gender & Society* 4(2):220–229 (June 1990).

Jassem, Harvey, and Theodore L. Glasser, "Children, Indecency, and the Perils of Broadcasting: The 'Scared Straight' Case," *Journalism Quarterly* 60(3):509–512 (Autumn 1983).

Jones, Steve, "Ban(ned) in the USA: Popular Music and Censorship," *Journal of Communication Inquiry* 15(1):73–87 (Winter 1991).

Kleiman, Howard M., "Indecent Programming on Cable Television: Legal and Social Dimensions," *Journal of Broadcasting & Electronic Media* 30(3):275–294 (Summer 1986).

Kontas, Gretta, "Gender, Disparaging Humor and Aggression: Have We Come Far Enough to Laugh, Baby?" unpublished paper prepared for presentation to the Speech Communication Association, November 1990.

Layman, William K., "NOTE: Violent Pornography and the Obscenity Doctrine: The Road Not Taken," Volume 75 *Georgetown Law Journal* 1475 (April 1987).

Lipschultz, Jeremy Harris, "The Function of Audience and Community Feedback in Broadcast Indecency Complaints and Station Management Responses: A Comparative Case Study of WLUP-AM and KSJO-FM," *Communications and the Law* 15(2):17–42 (June 1993).

Lipschultz, Jeremy Harris, "Conceptual Problems of Broadcast Indecency Policy and Application," *Communications and the Law* 14(2):3–29 (June 1992).

Lipschultz, Jeremy Harris, "'Political Propaganda': The Supreme Court Decision in Meese v. Keene," *Communications and the Law* 11(4):25–44 (December 1989).

Maretz, Heidi Skuba, "Aural Sex: Has Congress Gone Too Far by Going All the Way with Dial-a-Porn?" *Hastings Communication/Entertainment Law Journal* 11:493 (1989).

Messaris, Paul, and Dennis Kerr, "TV-Related Mother-Child Interaction and Children's Perceptions of TV Characters," *Journalism Quarterly* 61(3):662–666 (Autumn 1984).

Passler, Richard G., "Comment: Regulation of Indecent Radio Broadcasts: George Carlin Revisited—What Does the Future Hold for the Seven 'Dirty' Words?" Volume 65 *Tulane Law Review* 131 (November 1990).

Reiss, Guy A., "New F.C.C. Standards on Indecency on the Air and the First Amendment: Offensive Obscenity or Profound Profanity," *Columbia-VLA Journal of Law & the Arts* 13:221 (1989).

Smith, Tom W., "The Polls—A Report, The Sexual Revolution?" *Public Opinion Quarterly* 54:415–435 (1990).

Spitzer, Matthew L., "Controlling the Content of Print and Broadcasting," Volume 58 *Southern California Law Review* 1349 (1985).

Turow, Joseph, "Talk Show Radio as Interpersonal Communication," *Journal of Broadcasting* 18(2):171–179 (Spring 1974).

Winick, Charles, "The Social Contexts of Humor," *Journal of Communication* 26(3):124–128 (Summer 1976).

Zemach, Temera, and Akiba A. Cohen, "Perception of Gender Equality on Television and in Social Reality," *Journal of Broadcasting & Electronic Media* 30(4):427–444 (Fall 1986).

Books

Berger, Peter L., and Thomas Luckmann. *The Social Construction of Reality: A Treatise in the Sociology of Knowledge.* New York: Anchor, Doubleday, 1967.

Berman, Ronald. *Advertising and Social Change.* Beverly Hills: Sage, 1981.

Bittner, John R. *Law and Regulation of Electronic Media,* 2nd ed. Englewood Cliffs, N.J.: Prentice Hall, 1994.

Blumer, Herbert. *Industrialization As an Agent of Social Change: A Critical Analysis.* New York: Aldine de Gruyter, 1990.

Calhoun, Craig. "Indirect Relationships and Imagined Communities: Large-Scale Social Integration and the Transformation of Everyday Life," in *Social*

Theory for a Changing Society. Edited by Pierre Boudieu and James S. Coleman. Boulder: Westview Press, 1991.

Caristi, Dom. *Expanding Free Expression in the Marketplace: Broadcasting and the Public Forum.* New York: Quorum Books, 1992.

Carter, T. Barton, Marc A. Franklin, and Jay B. Wright. *The First Amendment and the Fifth Estate: Regulation of Electronic Mass Media,* 3rd ed. Westbury, N.Y.: The Foundation Press, 1993.

Carter, T. Barton, Marc A. Franklin, and Jay B. Wright. *The First Amendment and the Fifth Estate: Regulation of Electronic Mass Media,* 2nd ed. Westbury, N.Y.: The Foundation Press, 1989.

Chafee, Zechariah. "Government and Mass Communications," in *Reader in Public Opinion and Communication,* 2nd ed. Edited by Bernard Berelson and Morris Janowitz, New York: The Free Press, 1966.

Cohen, Akiba A., Hanna Adoni, and Charles R. Bantz. *Social Conflict and Television News.* Sage Library of Social Research 183, Newbury Park, Calif.: Sage, 1990.

Cohen, Jeremy, and Timothy Gleason. *Social Research in Communication and Law.* The Sage CommText Series, Vol. 23, Newbury Park, Calif.: Sage, 1990.

Comparatively Speaking: Communication and Culture Across Space and Time. Jay G. Blumler, Jack M. McLeod, and Karl Erik Rosengren, eds. Newbury Park, Calif.: Sage, 1992.

Comstock, George. *The Evolution of American Television.* Newbury Park, Calif.: Sage, 1989.

Creech, Kenneth C. *Electronic Media Law and Regulation.* Boston: Focal Press, 1993.

Dancy, Jonathan. *Moral Reasons.* Oxford: Blackwell, 1993.

Dominick, Joseph, Barry L. Sherman, and Gary Copeland. *Broadcasting/Cable and Beyond: An Introduction to Modern Electronic Media.* New York: McGraw-Hill, 1990.

Dominick, Joseph R., and James E. Fletcher. *Broadcasting Research Methods.* Newton, Mass.: Allyn and Bacon, 1985.

Emord, Jonathan W. *Freedom, Technology, and the First Amendment.* San Francisco: Pacific Research Institute for Public Policy, 1991.

Gillmor, Donald M., Jerome A. Barron, Todd F. Simon, and Herbert A. Terry. *Mass Communication Law: Cases and Comment,* 5th ed. St. Paul, Minn.: West Publishing, 1990.

Ginsburg, Douglas H., Michael H. Botein, and Mark D. Director. *Regulation of the Electronic Mass Media: Law and Policy for Radio, Television, Cable and the New Video Technologies,* 2nd ed. St. Paul, Minn.: West Publishing, 1991.

Goffman, Erving. "Presentation of Self in Everyday Life," in *Introducing Sociology: A Collection of Readings.* Edited by Richard T. Schaefer and Robert P. Lamm. New York: McGraw-Hill, 1987.

Graber, Doris A. *Mass Media and American Politics,* 3rd ed. Washington, D.C.: Congressional Quarterly, 1989.

Gray, Cavender. "'Scared Straight': Ideology and the Media," in *Justice and the Media: Issues and Research.* Edited by Ray Surette. Springfield, Ill.: Charles C. Thomas, 1984.

Greenwalt, Kent. *Law and Objectivity.* New York: Oxford University Press, 1992.

Hilliard, Robert L. *The Federal Communications Commission: A Primer.* Boston: Focal Press, 1991.

Holsti, Ole R. *Content Analysis for the Social Sciences and Humanities.* Reading, Mass.: Addison-Wesley, 1969.

Horton, Donald, and R. Richard Wohl. "Mass Communication and Parasocial Interaction: Observation on Intimacy at a Distance," in *InterMedia: Interpersonal Communication in a Media World,* 3rd ed. Edited by Gary Gumpert and Robert Cathcart, New York: Oxford University Press, 1986.

Kipnis, Kenneth. *Philosophical Issues in Law: Cases and Materials.* Englewood Cliffs: Prentice-Hall, 1977.

Krasnow, Erwin G., Lawrence D. Longley, and Herbert A. Terry. *The Politics of Broadcast Regulation,* 3rd ed. New York: St. Martin's Press, 1982.

Lowery, Shearon A., and Melvin L. DeFluer. *Milestones in Mass Communication,* 2nd ed. New York: Longman, 1988.

Mass Media and Society. Edited by James Curran and Michael Gurevitch. New York: Routledge and Chapman and Hall, 1991.

McQuail, Denis. *Mass Communication Theory: An Introduction,* 3rd ed. London: Sage, 1994.

Middleton, Kent R., and Bill F. Chamberlin. *The Law of Public Communication,* 3rd ed. New York: Longman, 1994.

Pember, Don R. *Mass Media Law,* 6th ed. Dubuque, Iowa: Brown & Benchmark, 1993.

Public Interest and the Business of Broadcasting: The Broadcast Industry Looks at Itself. Edited by Jon T. Powell and Wally Gair. New York: Quorum Books, 1988.

Ray, William B. *FCC: The Ups and Downs of Radio-TV Regulation.* Ames, IA: Iowa State University Press, 1990.

Schulz, Muriel R. "The Semantic Derogation of Woman," in *Language and Sex, Difference and Dominance.* Edited by Barrie Thorne and Nancy Henley. Rowley, Mass.: Newbury House, 1975.

Severin, Werner J., and James W. Tankard, Jr. *Communication Theories: Origins, Methods, and Uses in the Mass Media,* 3rd ed. New York: Longman, 1992.

Sherman, Barry L. *Telecommunications Management: The Broadcast & Cable Industries.* New York: McGraw-Hill, 1987.

Siebert, Fred S., Theodore Peterson, and Wilbur Schramm. *Four Theories of the Press: The Authoritarian, Libertarian, Social Responsibility and Soviet Communist Concepts of What the Press Should Be and Do.* Urbana: University of Illinois Press, 1963.

Smiley, Marion. *Moral Responsibility and the Boundaries of Community: Power and Accountability from a Pragmatic Point of View.* Chicago: The University of Chicago Press, 1992.

Spitzer, Matthew L. *Seven Dirty Words and Six Other Stories: Controlling the Content of Print and Broadcast.* New Haven: Yale University Press, 1986.

Stark, Werner. *The Sociology of Knowledge: Toward a Deeper Understanding of the History of Ideas.* New Brunswick, N.J.: Transaction Publishers, 1991.

Sterling, Christopher H., and John M. Kittross. *Stay Tuned: A Concise History of American Broadcasting,* 2nd ed. Belmont, Calif.: Wadsworth, 1990.

Stewart, Lea P., Alan D. Stewart, Sheryl A. Friendly, and Pamela J. Cooper. *Communication Between the Sexes: Sex Differences and Sex-Role Stereotypes,* 2nd ed. Scottsdale, Ariz.: Gorsuch Scarisbrick, 1990.

Thompson, Jr., Edward H., and Joseph H. Pleck. "The Structure of Male Role Norms," in *Changing Men: New Directions in Research on Men and Masculinity.* Edited by Michael S. Kimmel, Newbury Park, CA: Sage, 1987.

White, Cindy L. "Liberating Laughter: An Inquiry into the Nature, Content and Functions of Feminist Humor," in *Women Communicating: Studies of Women's Talk.* Edited by Barbara Bate and Anita Taylor, Norwood, N.J.: Ablex, 1988.

Wright, Charles R. *Mass Communication: A Sociological Perspective,* 3rd ed. New York: Random House, 1986.

Wright, R. George. *The Future of Free Speech Law.* New York: Quorum Books, 1990.

Cases

Abrams v. United States, 250 U.S. 616 (1919).

Action for Children's Television v. FCC (ACT III), 11 F.3d 170 (D.C. Cir. 1993).

Action for Children's Television v. FCC (ACT II), 932 F.2d 1504 (D.C. Cir. 1991).

Action for Children's Television v. FCC (ACT I), 271 U.S. App. D.C. 365, 852 F.2d 1332, 1340 (D.C. 1988).

Branton v. FCC, 993 F.2d 906 (D.C. Cir. 1993), cert. den.

Federal Communications Commission v. Pacifica Foundation, 438 U.S. 726 (1978).

Illinois Citizens Committee for Broadcasting v. FCC, 515 F. 2d 397 (D.C. Cir. 1974).

In Re Apparent Liability, WGLD-FM, 41 F.C.C. 2d 919 (1973).

In Re Infinity Broadcasting Corp. of Pa., 3 F.C.C.R. 930 (1987).

In Re WUHY-FM Eastern Educational Radio, 24 FCC 2d 408 (1970).

In the Matter of Liability of Sagittarius Broadcasting Corporation, 1992 FCC LEXIS 6042, (October 23, 1992).

In the Matter of Rob Warden, 70 F.C.C. 2d 1735 (1978).

Infinity Broadcasting Corp. of Pa., 2 FCC Rcd 2705, 3 FCC Rcd 930 (1987).

KING-TV, FCC 90–104, 1990 FCC LEXIS 2414, 10 May 1990.

Media Access Project, 41 F.C.C. 2d 179 (1973).

Miller v. California, 413 U.S. 15 (1973).

National Broadcasting Company v. United States, 319 U.S. 190 (1943).

Office of Communication of United Church of Christ v. FCC, 359 F.2d 994, 1005–1006 (D.C. Cir. 1966).

Order on reconsideration, Infinity Broadcasting Corp., 3 F.C.C. Rcd 930 (1987).

Pacifica Foundation, 36 FCC 147 (1964).

Pacifica Foundation, 2 F.C.C. Rcd. 2698 (1987).

R. v. Hicklin, L.R. 3 Q.B. 360 (1868).

Red Lion Broadcasting v. FCC, 395 U.S. 367 (1969).

Regents of the University of California, 2 F.C.C. Rcd. 2703 (1987).

Roth v. United States, 354 U.S. 476 (1957).

Telecommunications Research and Action Center v. FCC, 800 F. 2d 1181 (D.C. Cir. 1986).

Sonderling, 41 F.C.C. 2d 777 (1973).
United States v. O'Brien, 391 U.S. 367 (1968).
Yale Broadcasting v. FCC, 478 F.2d 594 (D.C. Cir. 1973).

Conference Papers

Driscoll, Paul D., "The Federal Communications Commission and Broadcast Indecency," Law Division, Association for Education in Journalism and Mass Communication, April 1989.

Glasser, Theodore L., "The Press, Privacy, and Community Mores," Mass Communication Division, Speech Communication Association, Louisville, Kentucky, November 1982.

Hindmarsh, Mike, "'How Is Pornographic?' (Not 'What Is Pornography?'),"unpublished paper prepared for presentation to the Communication Theory and Methodology Division, Association for Education in Journalism and Mass Communication, Minneapolis, August 1990.

Kim, Haeryon, "The Politics of Broadcast Deregulation: Beyond Krasnow, Longley, and Terry's 'Broadcast Policy-Making System'," unpublished paper prepared for presentation to the Association for Education in Journalism and Mass Communication, Minneapolis, August 1990.

Lipschultz, Jeremy Harris, "A Content Analysis of Broadcast Indecency Nonactionable Material," poster session paper, Broadcast Education Association, Radio `91, San Francisco, September 1991.

Shields, Steven O., "Creativity and Control in the Work of American Music Radio Announcers," unpublished paper prepared for presentation to the Mass Communication and Society Division, Association for Education in Journalism and Mass Communication, Minneapolis, August 1990.

Starr, Michael, and David Atkin, "The Department of Communications: A Plan for the Abolition of the Federal Communications Commission," Radio-Television Journalism Division, Association for Education in Journalism and Mass Communication, national conference, Washington, D.C., August 1989.

Terlip, Laura A., "Tough, Tender & Too Good to be True?: Student Attributions of Sex Roles to Successful Females in Situation Comedies," unpublished paper prepared for presentation to the Women's Caucus, Speech Communication Association, November 1990.

Laws

Amendment I (1791).
Broadcast Indecency in 18 U.S.C., Section 1464, 8 F.C.C.R. 704 (1993).
Freedom of Information Act, 5 U.S.C. Section 552.
In the Matter of Enforcement of Prohibitions Against Broadcast Obscenity and Indecency in 18 U.S.C. Sec. 1464, Order, FCC 88–416 (Dec. 19, 1988).
New Indecency Enforcement Standards, 2 F.C.C. Rcd. 2726 (1987).
47 United States Code, Section 326.
47 U.S.C., Sections 312 and 503.

Newspaper and Professional Journal Articles

Broadcasting, "FCC Reviews Sex, Wires and Video Topics," 27 January 1992, p. 14.

Broadcasting, "Quello Lauds 'Marketplace' Curbs on Indecency; Says Broadcasters, Advertisers Should Listen to Citizen Groups or Face Government Action," 27 January 1992, p. 39.

Broadcasting, Special Report: Station Trading 1991, 10 February 1992, p. 26.

Broadcasting, "Indecency Ban Comes under Fire; Appeals Court Judge Criticizes FCC's 24-Hour Indecency Prohibition, Challenging Commission's Contention that It Is 'Narrowly Tailored,'" 4 February 1991, pp. 40–41.

Broadcasting, "Life as a Washington Monument," 31 December 1990, p. 55.

Broadcasting, "Infinity to Fight FCC Indecency Fine," 3 December 1990, p. 38.

Broadcasting, "Stern Reprimanded," 5 November 1990, p. 6.

Broadcasting, "FCC Pulls Chicago TV License," 24 September 1990, p. 30.

Broadcasting, "The Ready Regulator," 13 August 1990, p. 28.

Broadcasting, "WGBH-TV Draws FCC Complaints, FCC Receives Letters about Mapplethorpe News Piece," 13 August 1990, p. 33.

Broadcasting, "FCC Votes 5-0 for Indecency Ban," 16 July 1990, p. 30.

Broadcasting, Law and Regulation column, "FCC Hears Little Support for 24-Hour Broadcasting Indecency Ban," 26 February 1990.

Broadcasting, "Radio Broadcasters Troubled by Sikes FCC 'Moving the Goal Posts' on Indecency," 6 November 1989, pp. 66–67.

Broadcasting, editorial, "Bad to Worse," 30 October 1989.

Broadcasting, "FCC Cleans out the Pipeline on Indecency," 30 October 1989, p. 28.

Broadcasting, "Chances Slim for 24-Hour FCC Ban on Indecency," 30 January 1989.

Buckley, Jr., William F., "Howard Stern Stumps as Libertarian," Universal Press Syndicate, *Omaha World-Herald*, June 1, 1994, p. 11.

Flint, Joe, "Indecency Rules under Fire in Courts, at FCC," *Broadcasting & Cable*, 1 March 1993, pp. 44–45.

Flint, Joe, "Hounding Howard: FCC's $100K Fine, Radio Personality Howard Stern," *Broadcasting & Cable*, 26 October 1992, p. 8.

Flint, Joe, "Evergreen to Fight Indecency Charge: Since It Has No Avenue of Appeal for FCC Fine, It Will Refuse to Pay; Matter Then Gets Handed over to Justice Department," *Broadcasting*, 13 January 1992, p. 91.

Graves, Rick, San Jose, Letters to the Editor, "Stone Defames the Ethnic Majority," *San Jose Mercury News*, 18 October 1988.

Halevi, Charles Chi, "Anti-Semitic Jokes on Radio, TV Aren't One Bit Funny," Commentary, *Chicago Sun-Times*, 25 March 1988, p. 44.

Hartman, Mitchell, "Attempts to Limit 'Indecent' Speech Worry Broadcasters," *The Quill*, October 1990, p. 32.

Holland, Bill, "Infinity License in Danger of Revocation over Stern?" *Billboard*, 10 April 1993, p. 69.

Kening, Dan, "Stern's Shock-Talk Show Not Playing in Chicago," *Chicago Tribune*, (zone n), 21 April 1993, p. 1.

Muller, Gale D., "WXRX radio special listenership study," The Gallup Organization, Princeton, New Jersey, December 1989.

National Association of Broadcasters, "FCC's Quello, Duggan Talk about AM Radio, Indecency," *Radio Week*, 19 March 1990, p. 3.

National Association of Broadcasters, *TV Today*, 26 February 1990.

Puig, Claudia, "Howard Stern: The Next Generation of Talk Radio; Industry Insiders Predict that the Nationwide Popularity of the Renegade Morning Man Will Spawn a Host of Imitators," *Los Angeles Times*, 8 October 1992, p. F1.

Radio-Television News Directors Association, "RTNDA Joins Appeal for Review of Indecency Rule," *Intercom* 6(2):2 (19 January 1989).

Ralston, Alice, San Jose, Letters to the Editor, "Stone's Irreverence Lightens Daily Load," *San Jose Mercury News*, 18 October 1988.

Shepherd, Chuck, *Minneapolis Star Tribune*, 15 April 1993, Metro Edition, p. 2E.

Steel, William, Redwood Estates, Letters to the Editor, "Protest San Jose's Racist Rock Station," *San Jose Mercury News*, 17 October 1988.

"Stern Quits Race in N.Y. to Keep Finances a Secret," Associated Press, *Omaha World-Herald*, 4 August 1994, p. 46.

"Stern Radio Show to be on Cable TV," Associated Press, *Omaha World-Herald*, 1 June 1994, p. 40.

USA Weekend, 25 April 1993, p. 14.

Wharton, Dennis, *Daily Variety*, 28 April 1993, p. 3.

Yourke, Jeffrey, "Stern Talk Results in Fine," *Washington Post*, 27 October 1992, p. C7.

Yourke, Jeffrey, "Locking on the Shock Jocks," *Washington Post*, 18 August 1992, p. D7.

Index